Collins

Collins Revision

GCSE Higher
Maths

Revision Guide

FOR EDEXCEL A
FOR AQA A
FOR AQA B

Keith Gordon

Contents

CONTENTS

Algebra

Statistics 1

Averages

- The **mode** is the **most common** value in a set of data.
- The **median** is the **middle number** when the data items are arranged in order.
- The **mean** of a set of data is the sum of all the values in the set divided by the total number of values in the set.
- The mean is defined as:

$$\text{mean} = \frac{\text{sum of all values}}{\text{total number of values}}$$

- When people talk about 'the average', they are usually referring to the mean.

D-C

Frequency tables

- When a lot of data has to be represented, it can be put into a **frequency table**.

This table shows how many times students in a form were late in a week.

Number of times late	0	1	2	3	4	5
Frequency	11	8	3	3	3	2

Top Tip!
The mode is the value that has the highest frequency; it is *not* the frequency itself.

- Find the **mode** by looking for the data value that has the **highest frequency**.
- Find the **median** by adding up the frequencies of the data items, in order, until the half-way point of all the data in the set is passed.
- Find the **mean** by multiplying the value of each data item by its frequency, adding the totals, then dividing by the total of all the frequencies.

Top Tip!
In examinations, data is usually given in vertical tables. Add another column to work out $x \times f$ (value multiplied by the frequency).

To find the average number of times students were late from the table above, work out the total number of times students were late as:
$0 \times 11 + 1 \times 8 + 2 \times 3 + 3 \times 3 + 4 \times 3 + 5 \times 2$
and work out the total frequency as: $11 + 8 + 3 + 3 + 3 + 2$,
then divide the first total by the second.

D-C

Questions

Grade D

1 Without using a calculator, find the mean, mode and median of these sets of data.

 a 8, 9, 4, 7, 8, 4, 9, 6, 3, 8

 b 11, 12, 10, 12, 15, 13, 11, 10, 11, 13, 14

Grade D

2 The number of times students in a form were late for school in a week is shown in the table above.

 a How many students were there altogether in the form?

 b What is the modal number of times students were late?

 c What is the median number of times students were late?

 d i What is the total number of times students were late?

  ii What is the mean number of times students were late per week?

C

Grouped data and frequency diagrams

- When there is a wide range of data, with lots of values, there are often too many entries for a frequency table. In this case, use a **grouped frequency table**.
- In a grouped frequency table, data is **recorded in groups** such as $10 < x \le 20$.
- The notation $10 < x \le 20$ means values between 10 and 20, not including 10 but including 20.
- Grouped data can be shown in a **frequency polygon**.
- In a frequency polygon, the **midpoint** of each group is plotted against the **frequency**.

This table shows the marks in a mathematics examination for 50 students. The frequency polygon shows the data.

Marks, x	Frequency, f
$0 < x \le 10$	4
$10 < x \le 20$	9
$20 < x \le 30$	17
$30 < x \le 40$	13
$40 < x \le 50$	7

Top Tip!

If asked to calculate an estimate for the mean, add two columns to the table: the first for the midpoint, m, and the second for $m \times f$.

- The **modal class** is the group with the greatest frequency.
- The **median** cannot be found from a grouped frequency table.
- Calculate an **estimate of the mean** by adding the midpoints multiplied by the frequencies and dividing by the total frequency.

A–A*

Histograms

- A **histogram** is similar to a bar chart, but is for continuous data only, such as time or weight.
- The **horizontal axis** has a **continuous** scale and there are no gaps between the bars.
- The **area** of each bar **represents** the class **frequency** of the bar.
- The height of the bars is called the **frequency density** and is calculated by:

$$\text{frequency density} = \frac{\text{frequency of class interval}}{\text{width of class interval}}$$

- The median is the value where the area on both sides is equal.
- The lower quartile is the value that splits the area in the ratio $1:3$.
- The upper quartile is the value that splits the area in the ratio $3:1$.

This histogram shows the masses of 100 snails.

Questions

(Grade C)

1 The marks for 50 students in a mathematics examination are shown in the table above.
  a What is the modal class?
  b i What is the total of the 'midpoints $\times$ frequencies'?
    ii What is the estimated mean mark for the form?

(Grade A)

2 Refer to the histogram above.
  a Work out how many snails had a mass between 5 and 10 grams.
  b Explain why the lower quartile is 15 grams.
  c What is the median mass of the snails?

Remember: You must revise all content from Grade E to the level that you are currently working at.

Moving averages

B–A

- A **moving average** gives a clear indication of the trend of a set of data. It **smooths out** data that has **seasonal variations**, such as the use of electricity for heating.

- A **two-point** moving average uses **two consecutive values** to calculate the average. A **three-point** moving average uses **three consecutive values**, and so on.

- To plot moving averages on a graph, plot each mean value at the **midpoint** of the **first** and **last** values used to calculate that average. The **trend** of the averages can be used to predict future values by using a line of best fit or the pattern of the data.

This table shows how many vans were hired from a rental company in a year.

Months	Jan	Feb	Mar	Apr	May	Jun	July	Aug	Sep	Oct	Nov	Dec
Vans	9	22	37	14	18	24	42	17	20	27	48	20

The four-point moving average for Jan–Apr is:
$(9 + 22 + 37 + 14) \div 4 = 20.5$.

The graph shows the data from the table plotted with the four-point moving averages.

Top Tip!

Draw the line of best fit on your diagram as examiners will 'follow through' your values to award marks.

Surveys

D-C

- **A survey** is an organised way of finding people's opinions or testing an hypothesis.
- Data from a survey is usually collected on a **data collection sheet**.
- Questionnaires are used to collect a lot of data.
- Questions on questionnaires should follow some rules.
  - Never ask a leading question
  - Never ask a personal question
  - Keep questions simple, with a few responses
  - Make sure the responses cover all possibilities and do not include any overlapping responses

Top Tip!

When you are asked to criticise questions in an examination, keep your answers short and simple, for example, 'It is a leading question'.

Questions

Grade B

1 a The graph above shows the number of vans hired from a rental company in a year and the four-point moving averages. Use this graph to predict the next moving average. Call it z.

b If the number of vans hired the following January is x, explain why $(27 + 48 + 20 + x) \div 4 = z$.

c Work out the value of x.

Grade D

2 The following are two questions used in a survey about recycling.
Give one reason why each question is not a good one.

a Recycling is a waste of time and does not help the environment.
Don't you agree? ☐ Yes ☐ No

b How many times a month do you use a bottle bank?
☐ Never ☐ 2 times or less ☐ More than 4 times

You must revise all content from Grade E to the level that you are currently working at.

Social statistics

D–C

- Social statistics is concerned with real-life statistics, such as:
  - The **Retail Price Index** (RPI) – one year is chosen as a reference year and prices for subsequent years are compared to this, with the change usually given as a percentage
  - **Time series** – these are similar to line graphs and they plot the changes over time of such things as employment rates or exchange rates between the pound and the dollar
  - The **national census** – the government in Britain takes a national census every ten years so they can keep track of changes in populations

Sampling

C–A

- Statisticians often have to **collect information** or **test hypotheses** about **a population**.
- In statistics, a **population** can be **any group** of people, objects or events.
- **Sampling** is a method of finding information, using a small sample from a large population.
- A sample needs to be:
  - **representative**: covering all the different groups within a population without bias
  - **of sufficient size**: large enough to make sure the results are valid for the whole population
- There are **two main types** of sample:
  - **random**: all members of the population have an equal chance of being chosen
  - **stratified**: the population is divided into categories and a number of each category is surveyed in the same proportion as in the population. The sample within each category is taken randomly.

This table shows the numbers of students in each year of a school.

School year	Boys	Girls	Total
7	52	68	120
8	46	51	97
9	62	59	121
10	47	61	108
11	39	55	94
Total number in school			540

Top Tip!

When asked to comment on sampling methods in an examination, keep your answers short and simple such as, 'It is not a random process'.

Questions

Grade C

1 The euro was introduced in January 2002. Initially the exchange rate between the euro and the pound was €1.65 = £1. In January 2007, the exchange rate was €1.40 = £1. Taking January 2002 as the reference year with an exchange rate index of 100, what is the exchange rate index for January 2007? Give your answer to the nearest whole number.

Grade A

2 a The table above shows the numbers of students in each year of a school. The headteacher wants to survey 50 of these students. Which of the three methods below would give a random sample?

Explain your choice and give a reason why the other methods would not be suitable.
  i Asking two year 7 forms of 25 students
  ii Asking 50 students on the first bus in the morning
  iii Using an alphabetical list of all the students, assigning a number to each student, putting 540 raffle tickets in a hat, then picking out 50 tickets

  b The deputy head decides to take a stratified sample of 10% of the school. How many students from each year group should she survey?

Statistics 2

Line graphs

D–C

- **Line graphs** are used to show how data changes over a period of time.

- Line graphs can be used to show **trends**, such as how the average daily temperature changes over the year.

- **Data points** on line graphs can be joined by **lines**.

  – When the lines join points that show **continuous data**, such as points showing the height of a plant each day over a week, they are drawn as **solid lines** because the lines can be used to estimate intermediate values.

  – When the lines join points that show discrete data, such as the average daily temperature, they are drawn as **dotted lines**, because the lines cannot be used to estimate intermediate values.

This line graph shows the temperatures in a town in Australia and a town in Britain.

> **Top Tip!**
> The largest increase or decrease on a line graph is at the steepest part of the graph.

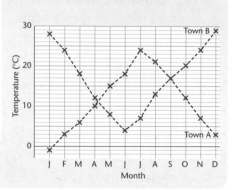

Temperatures in a town in Australia and a town in Britain

Stem-and-leaf diagrams

D–C

- When data is first recorded, it is called **raw data** and is **unordered**.

- Unordered data can be put into order to make it easier to read and understand. This is called **ordered** data.

- A **stem-and-leaf diagram** is a way of showing ordered data.

- The **stem** is the number on the left of the vertical line. The **leaves** are the numbers on the right of the vertical line.

- The **mode** is the most common entry.

- You can find the **median** by counting from the start to the middle value.

- You can find the **range** by subtracting the lowest value from the highest value.

- You can find the **mean** by adding all the values and dividing by the total number of values in the table.

> **Top Tip!**
> Always make sure you add a key to show what the stem and leaves represent.

This stem-and-leaf diagram shows the marks scored by students in a test.

1	3 4 5 6
2	0 1 1 3 4 5 8
3	1 2 5 5 5 9
4	2 2 9

Key 1 | 3 represents 13 marks

Questions

Grade D

1 Refer to the line graph above.

  a Which town is hotter, on average? Give a reason for your answer.

  b Which town is in Australia? Give a reason for your answer.

  c In which month was the average temperature the same in both towns?

  d Is it true that the average temperature was the same in both towns on a day in early April?

Grade D

2 Refer to the stem-and-leaf diagram above.

  a How many students took the test?

  b What is the range of the marks? Show your working.

  c What is the modal mark?

  d What is the median mark?

  e What is the mean mark? Show your working.

Scatter diagrams

E–C

Scatter diagrams

- A **scatter diagram** (also known as a **scattergraph** or **scattergram**) is a diagram for comparing two variables.
- The variables are plotted as **coordinates**.

Here are 10 students' marks for two tests.

Tables	3	7	8	4	6	3	9	10	8	6
Spelling	4	6	7	5	5	3	10	10	9	7

This is the scatter diagram for these marks.

E–C

Correlation

- The scatter diagram will show a **relationship** between the variables if there is one.
- The relationship is described as **correlation** and can be written as a **'real-life' statement**.

For the first diagram: 'The taller people are, the bigger their arm span'.

E–C

Line of best fit

- A **line of best fit** can be drawn through the data.
- The line of best fit can be used to **predict** the value of one variable when the other is known.

Top Tip!

Draw the line of best fit between the points with about the same number of points on either side of it.

The line of best fit passes through the 'middle' of the data.

Questions

Grade D

1 a Refer to the scatter diagram of students' scores for tables and spelling tests, above. What type of correlation does the scatter diagram show?

b Describe the relationship in words.

Grade C

2 a Describe, in words, the relationship between the value of a car and the age of the car, shown in the second of the three scatter diagrams above.

b Describe, in words, the relationship between the wages and the distance travelled to work, shown in the third of the three scatter diagrams above.

Grade C

3 Refer to the line of best fit on the scatter diagram showing the English and mathematics marks. Estimate the score in the English examination for someone who scored 75 in the mathematics examination.

Remember: You must revise all content from Grade E to the level that you are currently working at.

Cumulative frequency diagrams

B

- The **interquartile range** is a measure of the **dispersion** of a set of data.

- The interquartile range **eliminates extreme values** and bases the measure of spread on the middle 50% of the data.

- The interquartile range and the median can be found by drawing a cumulative frequency diagram. Add the frequency to the sum of the preceding frequencies to find the cumulative frequency.

This grouped table shows the marks of 50 students in a mathematics test and the cumulative frequencies.

The graph shows the cumulative frequency diagram for the same data.

Mark	Number of students	Cumulative frequency
21 to 40	7	7
41 to 60	14	21
61 to 80	14	35
81 to 100	13	48
101 to 120	2	50

- The points of a cumulative frequency diagram are plotted as the **top value** of each group against its **cumulative frequency**.

- To find the **lower quartile**, **median** and **upper quartile**, draw horizontal lines from the quarter, half and three-quarter values to the graph, then read the values from the horizontal axis.

- To find the **interquartile range**, subtract the lower quartile from the upper quartile.

Box plots

B

Top Tip!
Box plots are always plotted against a **scale** so that their values can be accurately read and plotted.

- Another way of displaying data for comparison is by means of a **box-and-whisker plot** (or just **box plot**).

- Box plots can be used to show the spread of two or more sets of data and make it easier to compare data.

- A box plot requires five pieces of data: the **lowest value** of the set of data, the **lower quartile** (Q_1) of the set of data, the **median** (Q_2) of the set of data, the **upper quartile** (Q_3) of the set of data and the **highest value** of the set of data.

Lowest value   Lower quartile, Q_1 Median, Q_2 Upper quartile, Q_3 Highest value

Questions

Grade B

1 Refer to the cumulative frequency diagram above.
  a What is the median score?
  b What is the lower quartile?
  c What is the upper quartile?
  d What is the interquartile range?
  e The lowest score is 21. The highest score is 120. Draw a box plot to show the data using a scale from 20 to 120.

Probability

Relative frequency

- The probability of an event can be found by carrying out many trials and calculating the **relative frequency** or **experimental probability**.

- To calculate the relative frequency of an event, **divide** the number of times the event occurred during the experiment by the **total number of trials** done in the experiment.

- The **higher the number of trials** carried out, the nearer the experimental probability will be to the true probability.

Top Tip!
If you are asked who has the best set of results, always say the person with the most trials.

These are the results when three students tested the spinner.

Student	Ali	Barry	Clarrie
Number of throws	20	60	240
Number of 4s	5	13	45

Mutually exclusive and exhaustive events

- When the outcome of event A can never happen at the same time as the outcome of event B, then event A and event B are said to be **mutually exclusive**.

  If a bag contains black and yellow balls, the events 'picking a yellow ball' and 'picking a black ball' can never happen at the same time when only one ball is taken out: that is, a ball can be either black or yellow.

- When the probability of mutually exclusive events add up to 1, they are called **exhaustive events**.

  Throwing an odd or an even number when throwing a dice is an exhaustive event because you will either throw an odd number or an even number.

- If there is an event A, the **complementary** event of A is event A *not* happening.

  The probability of drawing an ace from a pack of cards is $\frac{4}{52} = \frac{1}{13}$, so the probability of not drawing an ace is $1 - \frac{1}{13} = \frac{12}{13}$.

Questions

Grade C

1 The table above shows the results when three students tested a home-made spinner.

  a For each student, calculate the relative frequency of getting a 4. Give your answers to 2 decimal places.

  b Which student has the most reliable estimate of the actual probability of a 4? Give a reason for your answer.

  c If the spinner was fair, how many times would you expect it to land on 4 in 240 spins?

Grade C

2 a A card is taken at random from a pack of cards.
  i What is the probability it is red?
  ii What is the probability it is black?
  iii Explain why the events in parts **i** and **ii** are mutually exclusive.
  iv Explain why the events in parts **i** and **ii** are exhaustive.

  b A card is taken at random from a pack of cards.
  i What is the probability it is a king?
  ii What is the probability it is a **not** a king?

Remember: You must revise all content from Grade E to the level that you are currently working at.

Expectation

- When the probability of an event is known, you can predict how many times the event is likely to happen in a given number of trials.

- This is the **expectation**. It is *not* what is going to happen.

  If a coin were tossed 1000 times, the expectation would be 500 heads and 500 tails. It is very unlikely that this would actually occur in real life.

- The expected number is calculated as: **expected number = P(event) × total trials**

D

Two-way tables

- A **two-way table** is a table that links two variables.

- One of the **variables** is shown by the rows of the table.

- One of the **variables** is shown by the **columns** of the table.

D-C

This table shows the nationalities of people on a plane and the types of tickets they have.

	First class	Business class	Economy
American	6	8	51
British	3	5	73
French	0	4	34
German	1	3	12

Addition rule for events

- To work out the probability of either of two mutually exclusive events happening, **add** the probabilities of all the separate events: **P(A or B) = P(A) + P(B)**

D-C

Questions

Grade D

1 A bag contains ten counters. Five are red, three are blue and two are white. A counter is taken from the bag at random. The colour is noted and it is replaced in the bag. This is repeated 100 times.

  a How many times would you expect a red counter to be taken out?

  b How many times would you expect a white counter to be taken out?

  c How many times would you expect a red or a white counter to be taken out?

Grade D

2 Refer to the table above.

  a How many travellers were on the plane altogether?

  b What percentage of the travellers had first-class tickets?

  c What percentage of the business-class passengers was American?

Grade D

3 A card is taken at random from a pack of cards.

  a What is the probability that it is an ace?

  b What is the probability that it is a king?

  c What is the probability that it is an ace or a king?

  d What is the probability that it is not an ace?

  e What is the probability that it is not an ace or a king?

Remember: You must revise all content from Grade E to the level that you are currently working at.

Combined events

D-C

- When two events occur together they are known as **combined events**.
- The outcomes of the combined events can be shown as a list.

  If two coins are thrown together, the possible outcomes are (head, head), (head, tail), (tail, head) and (tail, tail).

- Another method for showing the outcomes of a combined event is to use a **sample space diagram**.

  If two dice are thrown, the outcomes can be shown as:

Outcomes

Score on red dice						
6	(1, 6)	(2, 6)	(3, 6)	(4, 6)	(5, 6)	(6, 6)
5	(1, 5)	(2, 5)	(3, 5)	(4, 5)	(5, 5)	(6, 5)
4	(1, 4)	(2, 4)	(3, 4)	(4, 4)	(5, 4)	(6, 4)
3	(1, 3)	(2, 3)	(3, 3)	(4, 3)	(5, 3)	(6, 3)
2	(1, 2)	(2, 2)	(3, 2)	(4, 2)	(5, 2)	(6, 2)
1	(1, 1)	(2, 1)	(3, 1)	(4, 1)	(5, 1)	(6, 1)
	1	2	3	4	5	6

Score on blue dice

Total score

Score on red dice						
6	7	8	9	10	11	12
5	6	7	8	9	10	11
4	5	6	7	8	9	10
3	4	5	6	7	8	9
2	3	4	5	6	7	8
1	2	3	4	5	6	7
	1	2	3	4	5	6

Score on blue dice

Tree diagrams

B

- An alternative method for showing the outcomes of combined events is a **tree diagram**.

  This tree diagram shows the outcomes of the combined events, tossing a coin followed by throwing a six with a dice.

- Note that the probabilities across any branch add up to 1.

- The probability of any outcome is calculated by multiplying together the probabilities on its branches.

First event	Second event	Outcome	Probability
$\frac{1}{2}$ head	$\frac{1}{6}$ six / $\frac{5}{6}$ not six	Head and Six	$\frac{1}{2} \times \frac{1}{6} = \frac{1}{12}$
$\frac{1}{2}$ tail	$\frac{1}{6}$ six / $\frac{5}{6}$ not six		

Top Tip!

In an examination, all or part of the tree diagram is always drawn and you just have to complete it.

The probability of a head followed by a 6 has been calculated in the above tree diagram.

Questions

Grade C

1 a When two dice are thrown together, how many possible outcomes are there?

  b Refer to the left-hand sample space diagram for throwing two dice, above.

    i What is the probability of throwing a double with two dice?

    ii What is the probability that the difference between the scores on the two dice is 4?

  c Refer to the right-hand sample space diagram for throwing two dice, above.

    i What is the probability of throwing a score of 5 with two dice?

    ii What is the probability of throwing a score greater than 9 with two dice?

    iii What is the most likely score with two dice?

Grade B

2 Using the tree diagram above, what is the probability of each of the following outcomes?

  a a head followed by not a six

  b a tail followed by a six

  c a tail followed by not a six

(Remember:) You must revise all content from Grade E to the level that you are currently working at.

Independent events

B–A

- If the outcome of event A cannot effect the outcome of event B, then events A and B are known as **independent events**.

- Most of the combined events we have revised so far with sample space diagrams and tree diagrams have been independent events.

- An alternative method for working out the outcomes of combined independent events is to use **and** and **or**.

Top Tip!

Notice that and is replaced with $\times$ **and or** is replaced with $+$.

On a fairground stall, Steven shoots darts at a target. The probability that he hits the target with any one dart is $\frac{2}{3}$. He has two darts. What is the probability that he:

a hits the target twice?

P(hits both times) = P(first dart hits **and** second dart hits)
= P(hit) $\times$ P(hit) = $\frac{2}{3} \times \frac{2}{3} = \frac{4}{9}$.

b hits the target once only?

P(hits the target once only) = P(first hits **and** second misses **or** first misses **and** second hits) = $\frac{2}{3} \times \frac{1}{3} + \frac{1}{3} \times \frac{2}{3} = \frac{2}{9} + \frac{2}{9} = \frac{4}{9}$.

At least problems

A

- In examination questions concerning combined events, it is common to ask for the probability of at least one of the events occurring.

- As we are dealing with independent events, **P(at least one) = 1 – P(none)**.

A fair dice is thrown three times. What is the probability of at least one 6 being thrown?

P(at least one 6) = $1 - $ P(no 6s) = $1 - \frac{5}{6} \times \frac{5}{6} \times \frac{5}{6} = 1 - \frac{125}{216} = \frac{91}{216}$.

Conditional probability

A*

- When the probability of an event is dependent on the outcome of another (preceding) event, it is known as conditional probability.

Top Tip!

A tree diagram can be used to solve conditional probability problems.

A drawer contain seven socks, three are red and four are blue. Two socks are taken out of the drawer at random. What is the probability that they are the same colour?

P(same colour) = P(red, red **or** blue, blue) = $\frac{3}{7} \times \frac{2}{6} + \frac{4}{7} \times \frac{3}{6} = \frac{3}{7}$.

Questions

Grade A

1 Annie throws a biased dice twice. The probability of throwing a six with the dice is $\frac{1}{4}$. What is the probability that Annie throws:

  a two sixes?

  b at least one six?

Grade A*

2 A box contains four white eggs and two brown eggs. Karen needs two eggs to bake a cake. She takes two eggs at random from the box. What is the probability that:

  a both eggs are brown?

  b both eggs are white?

  c one egg is white and one egg is brown?

Handling data checklist

I can...

☐ draw an ordered stem-and-leaf diagram

☐ find the mean of a frequency table of discrete data

☐ draw a frequency polygon for discrete data

☐ find the mean from a stem-and-leaf diagram

☐ predict the expected number of outcomes of an event

☐ draw a line of best fit on a scatter diagram

☐ recognise the different types of correlation

☐ design a data collection sheet

☐ use the total probability of 1 to calculate the probabilities of events

You are working at ⬭ Grade D ⬭ level.

☐ find an estimate of the mean from a grouped table of continuous data

☐ draw a frequency diagram for continuous data

☐ calculate the relative frequency of an event from experimental data

☐ interpret a scatter diagram

☐ use a line of best fit to predict values

☐ design and criticise questions for questionnaires

You are working at ⬭ Grade C ⬭ level.

☐ calculate an n-point moving average

☐ use a moving average to predict future values

☐ draw a cumulative frequency diagram

☐ find the median, quartiles and interquartile range from a cumulative frequency diagram

☐ draw and interpret box plots

☐ draw a tree diagram to work out probabilities of combined events

You are working at ⬭ Grade B ⬭ level.

☐ draw histograms from frequency tables with unequal intervals

☐ calculate the numbers to be sampled for a stratified sample

☐ use AND/OR to work out the probabilities of combined events

You are working at ⬭ Grade A ⬭ level.

☐ find the median, quartiles and interquartile range from a histogram

☐ work out the probabilities of combined events when the probabilities change depending on previous outcomes (conditional probability)

You are working at ⬭ Grade A* ⬭ level.

Number

Real-life problems

- In a GCSE examination, long multiplication and long division are usually given in the context of **real-life problems**.

- You will need to **identify** the problem as a **multiplication** or **division** and then work it out by your **preferred method**.

- In the GCSE non-calculator paper, questions often ask such things as 'How many coaches are needed?' and the calculation gives a remainder. Remember that an extra coach will be needed to carry the remaining passengers. You cannot have a 'bit of a coach'.

Top Tip!
Show your working clearly because even if you make a small arithmetical error, you will still get marks for method.

D

A café uses 950 eggs per day. The eggs come in trays of 24. How many trays of eggs will the café need?

This is a division problem. It is done using the traditional method.

$$\begin{array}{r} 39 \\ 24\overline{)950} \\ \underline{720} \\ 230 \\ \underline{216} \\ 14 \end{array}$$

The answer to the calculation

$950 \div 24$ is 39 remainder 14.

The café will need 40 trays and will have 10 eggs left over.

Work out 243×68.
Using the box method, split the two numbers into hundreds, tens and units, write them in a grid like the one below and multiply all the pairs.

×	200	40	3
60	12 000	2400	180
8	1 600	320	24

Add the separate answers to find the total.

So $243 \times 68 = 16\ 524$

Dividing by decimals

- To divide by a decimal, it is advisable to change the problem into one where you divide by an integer.

- The divisor and the dividend are multiplied by 10 or 100, etc.

Evaluate the following. **a** $34 \div 0.2$ **b** $16.2 \div 0.45$

a $34 \div 0.2 = 340 \div 2 = 170$

b $16.2 \div 0.45 = 1620 \div 45$, which then becomes a long division problem to which the answer is 36.

D

$$\begin{array}{r} 1620 \\ \underline{900} \quad 20 \times 45 \\ 720 \\ \underline{450} \quad 10 \times 45 \\ 270 \\ \underline{180} \quad\ \ 4 \times 45 \\ 90 \\ \underline{90} \quad\ \ 2 \times 45 \\ 0 \quad 36 \times 45 \end{array}$$

Questions

Grade D

1 a There are 945 students in a school. There are 27 students in each tutor group. How many tutor groups are there?

b To raise money for charity, one tutor group decides that all 27 members will donate the cost of a KitKat. If a KitKat costs 42p, how much money do they raise?

Grade D

2 Work out each of the following.

a $7.2 \div 0.4$

b $7.8 \div 0.6$

c $5.16 \div 0.43$

Remember: You must revise all content from Grade E to the level that you are currently working at.

D-C

Rounding and approximating

- To round a number to n significant figures, find the number that is closest with n digits and the rest of the places made up with zeros.

432 is 400 to 1sf and 430 to 2 sf.
0.08763 is 0.09 to 1sf, 0.088 to 2 sf and 0.0876 to 3 sf.

- To **approximate** the answer to a calculation, round each number in the calculation to 1 significant figure, then work out the calculation.

$38.2 \times 9.6 \approx 40 \times 10 = 400$, so $38.2 \times 9.6 \approx 400$
$48.3 \div 19.7 \approx 50 \div 20 = 2.5$, so $48.3 \div 19.7 \approx 2.5$

- The sign $\approx$ means 'is approximately equal to'.

Top Tip!
In your GCSE examination, you could be asked to round to 1, 2 or 3 significant figures. Final answers in exams should not be rounded to less than 3 significant figures unless you are instructed otherwise.

D

Multiplying and dividing by powers of 10

- When you **multiply** by a power of 10, the digits of the number move to the **left**.
- When you **divide** by a power of 10, the digits of the number move to the **right**.
- The number of **places the digits move** depend on the number of zeros or the power of 10.

Top Tip!
Although strictly speaking the digits move, it may be easier to think of it as the decimal point moving.

$45.9 \div 10^2 = 0.459$

10	1	$\frac{1}{10}$	$\frac{1}{100}$	$\frac{1}{1000}$	
4	5 . 9				$\div 10^2$
	. 4	5	9		

$45.9 \times 10^2 = 4590$

1000	100	10	1	$\frac{1}{10}$	
		4	5 . 9		$\times 10^2$
4	5	9	0 . 0		

D

Multiplying and dividing multiples of 10

- To **multiply** together two multiples of 10, multiply the non-zero digits and write the total of the zeros in both numbers at the end.

- To **divide** one multiple of 10 by another, divide the non-zero digits and write the difference in the zeros in both numbers at the end.

$200 \times 4000 = 800\ 000$
$500 \times 60 = 30\ 000$
$8000 \div 20 = 400$
$20\ 000 \div 40 = 500$

Questions

Grade C
1 a Round each of these numbers to 1 significant figure.
 i 3.8   ii 0.752   iii 58.7
 b Round each of these numbers to 2 significant figures.
 i 56.8   ii 0.965   iii 88.9

Grades D-C
2 Find an approximate answer to each of the following.
 a 68.3×12.2 b $203.7 \div 38.1$
 c $\dfrac{78.3 + 19.6}{21.8 - 9.8}$ d $\dfrac{42.1 + 78.6}{4.7 - 19.3}$

Grade D
3 Write down the answer to each of the following.
 a 0.3×10^3 b 7.6×100 c $0.75 \div 10$
 d $34 \div 10^3$ e 3000×200 f $4000 \div 20$

Prime factors

C

- When a number is written as a product of **prime factors**, it is written as a multiplication consisting only of prime numbers.

  $30 = 2 \times 3 \times 5 \qquad 50 = 2 \times 5 \times 5$ or 2×5^2

- To find the prime factors of a number, divide by prime numbers until the answer is a prime number.

Top Tip!

If a number is even, 2 is an obvious choice as a divisor or part of the product. Then, look at 3, 5 …

Find the prime factors of 24.
Divide by prime numbers until the answer is a prime number.
So $24 = 2 \times 2 \times 2 \times 3$

```
2 ) 2 4
2 ) 1 2
2 )   6
        3
```

- Products of prime factors can be expressed in **index form**.

  $24 = 2 \times 2 \times 2 \times 3 = 2^3 \times 3 \qquad 76 = 2 \times 2 \times 19 = 2^2 \times 19$

Lowest common multiple and highest common factor

C

- The **lowest common multiple** (**LCM**) of two numbers is the smallest number in the times tables of both of the numbers.

  The LCM of 6 and 7 is 42.　　The LCM of 8 and 20 is 40.

Top Tip!

Use the list method as this is much easier and gets full credit even if you have already done part of the prime factor method in an earlier part of the question.

- The **highest common factor** (**HCF**) of two numbers is the biggest number that divides exactly into the two numbers.

  The HCF of 24 and 18 is 6.　　The HCF of 45 and 36 is 9.

- To find the LCM and HCF, write out the multiples and factors of each number.

Find the LCM and HCF of 16 and 20.

LCM: Write out the 16 and 20 times tables, continuing until there is a common multiple.

16: 16, 32, 48, 64, 80, 96, … 20: 20, 40, 60, 80, 100, … So the LCM of 16 and 20 is 80.

HCF: Write out the factors of 16 and 20 then pick out the biggest number that appears in both lists.

Factors of 16: {1, 2, 4, 8, 16} Factors of 20: {1, 2, 4, 5, 10, 20} So the HCF of 16 and 20 is 4.

Questions

Grade C

1 a What numbers are represented by these products of prime factors?
  i $2 \times 3 \times 5$　**ii** $2 \times 2 \times 3 \times 7$　**iii** $2 \times 5 \times 13$
  iv $2^2 \times 3^2$　**v** $2^3 \times 5$　**vi** $2^2 \times 3 \times 5^2$

b Use the division method to find the prime factors of each of these numbers. Give your answers in index form where possible.
  i 20　**ii** 45　**iii** 64　**iv** 120

Grade C

2 a Find the LCM of the numbers in each pair.
  i 5 and 6　**ii** 3 and 7　**iii** 3 and 13

b Describe a connection between the LCM and the original numbers in part **a**.

c Find the LCM of each the numbers in each pair.
  i 6 and 9　**ii** 8 and 20　**iii** 15 and 25

d Find the HCF of each of the numbers in each pair.
  i 12 and 30　**ii** 18 and 40　**iii** 15 and 50
  iv 16 and 80　**v** 24 and 60　**vi** 12 and 25

Fractions

One quantity as a fraction of another

D

- To write one quantity as a fraction of another, write the **first quantity** as the **numerator** and the **second quantity** as the **denominator**.

What is £8 as a fraction of £20?

Write as $\frac{8}{20}$ then cancel to $\frac{2}{5}$.

Top Tip!

Examination questions often ask for the fraction to be given in its simplest form. This means it has to be cancelled down.

Adding and subtracting fractions

D-C

- When **adding** and **subtracting** fractions, use **equivalent fractions** to make the denominators of the fractions the same.

$$\frac{3}{8} + \frac{1}{5} = \frac{3 \times 5}{8 \times 5} + \frac{1 \times 8}{5 \times 8} = \frac{15 + 8}{40} = \frac{23}{40}$$

- When adding or subtracting mixed numbers, split the calculation into whole numbers and fractions.

$$4\frac{2}{5} - 1\frac{3}{4} = 4 - 1 + \frac{2}{5} - \frac{3}{4} = 3 + \frac{8}{20} - \frac{15}{20} = 3 + -\frac{7}{20} = 2\frac{13}{20}$$

Top Tip!

Notice that you have to split one of the whole numbers if the answer to the fractional part is negative.

Multiplying and dividing fractions

D-C

- To **multiply** two fractions, multiply the numerators and multiply the denominators.

$$\frac{3}{4} \times \frac{3}{7} = \frac{3 \times 3}{4 \times 7} = \frac{9}{28}$$

- When multiplying **mixed numbers**, change the mixed numbers into **top-heavy** fractions then multiply, as for ordinary fractions.

- **Cancel** any common factors in the top and bottom before multiplying.

- Convert the **final answer** back to a **mixed number** if necessary.

$$3\frac{1}{4} \times 1\frac{1}{5} = \frac{13}{4\!\!\!/_2} \times \frac{\cancel{6}^3}{5} = \frac{13 \times 3}{2 \times 5} = \frac{39}{10} = 3\frac{9}{10}$$

- To **divide** by a fraction, turn it upside down and multiply by it.

$$\frac{3}{8} \div \frac{7}{9} = \frac{3}{8} \times \frac{9}{7} = \frac{27}{56}$$

- When dividing mixed numbers, change them into top-heavy fractions and then divide, as for ordinary fractions.

When the second fraction has been turned upside down, cancel before multiplying.

$$1\frac{1}{4} \div 1\frac{7}{8} = \frac{5}{4} \div \frac{15}{8} = \frac{\cancel{5}^1}{\cancel{4}^1} \times \frac{\cancel{8}^2}{\cancel{15}^3} = \frac{2}{3}$$

Questions

Grade D

1 a What fraction of 25 is 10?

  b In a class of 28 students, 21 are right-handed. What fraction is this?

  c What fraction is 20 minutes of one hour?

Grades D-C

2 Work out each of these.

  a i $\frac{1}{4} + \frac{3}{7}$ **ii** $\frac{5}{6} + \frac{4}{9}$ **iii** $3\frac{2}{3} + 2\frac{2}{5}$

  b i $\frac{3}{5} - \frac{1}{6}$ **ii** $\frac{8}{9} - \frac{2}{3}$ **iii** $2\frac{1}{4} - 1\frac{2}{3}$

  c i $\frac{3}{4} \times \frac{2}{9}$ **ii** $\frac{5}{8} \times \frac{4}{7}$ **iii** $1\frac{2}{5} \times 2\frac{3}{4}$

  d i $\frac{3}{5} \div \frac{6}{7}$ **ii** $\frac{5}{6} \div \frac{10}{21}$ **iii** $3\frac{3}{5} \div 2\frac{1}{4}$

Percentage

The percentage multiplier

- Using the **percentage multiplier** is the best way to solve percentage problems.

- The percentage multiplier is the **percentage** expressed as a **decimal**.

  72% gives a multiplier of 0.72.   20% gives a multiplier of 0.20 or 0.2.

- The multiplier for a percentage **increase or decrease** is the percentage multiplier **added to or subtracted from 1**.

  An 8% increase is a multiplier of 1.08 (1 + 0.08).

  A 5% decrease is a multiplier of 0.95 (1 – 0.05).

Top Tip!

Learn how to use multipliers as they make percentage calculations easier and more accurate.

D-C

Calculating a percentage increase or decrease

- To calculate the new value after a quantity is **increased** or **decreased** by a percentage, simply **multiply** the original quantity by the **percentage multiplier** for the increase or decrease.

  What is the new cost after a price of £56 is decreased by 15%?

  Work it out as 0.85 × 56 = £47.60.

Top Tip!

If the calculator shows a number such as 47.6 as the answer to a money problem, always put the extra zero into the answer, so you write this down as £47.60.

D-C

Expressing one quantity as a percentage of another

- To calculate one **quantity as a percentage of another**, divide the first quantity by the second. This will give a decimal, which can be converted to a percentage.

  A plant grows from 30 cm to 39 cm in a week.

  What is the percentage growth?

  The increase is 9 cm. 9 ÷ 30 = 0.3 and this is 30%.

Top Tip!

Always divide by the original quantity, otherwise you will not get any marks.

C

Questions

Grades D-C

1 a Write down the percentage multiplier for each of these.
   i 80%   ii 7%   iii 22%

  b Write down the multiplier for each percentage increase.
   i 5%   ii 12%   iii 3.2%

  c Write down the multiplier for each percentage decrease.
   i 8%   ii 15%   iii 4%

Grades D-C

2 a Increase £150 by 12%.
  b Decrease 72 kg by 8%.

Grade C

3 a The average attendance at Barnsley football club in 2005 was 14 800.
   In 2006 it was 15 540.
   i By how much had the average attendance gone up?
   ii What is the percentage increase in attendance?

  b After her diet, Carol's weight had gone from 80 kg to 64 kg.
   What is the percentage decrease in her weight?

C

Compound interest

- When money is **invested** in, for example, a savings account, it can earn a certain rate of **interest** each year.

- This interest is then added to the original amount and the **new total amount** then earns interest at the same rate in the following year.

- The process whereby interest is **accumulated** each year is called **compound interest**.

- The best way to calculate the compound interest is to use a **multiplier**.

> A bank pays 4% compound interest per year. Jack invests £3000.
> How much will he have after 5 years?
> The multiplier for an annual increase of 4% is 1.04
> The calculation is: $3000 \times 1.04^5 = £3649.96$

- Remember the formula as **total amount = $P \times (1 + x)^n$**, where P is the original amount invested, x is the interest rate expressed as a decimal and n is the number of years for which the money is invested.

- This type of problem can also be about increasing or decreasing populations, salaries, weights, etc.

> A Petri dish containing 20 000 bacteria is treated with a detergent which kills 12% of the bacteria each minute. How many bacteria will remain after 10 minutes?
> The multiplier for a decrease of 12% is 0.88.
> The calculation is: $20\ 000 \times 0.88^{10} = 5570$

Top Tip!

If the answer is a string of decimals, round off to something sensible, that is a whole number for discrete data or 2 decimal places for money.

B

Reverse percentage

- After an amount has been increased or decreased by a certain percentage, the original amount can be found from the new amount.

- There are two methods to do this: the **unitary method** and the **multiplier method**.

> After a 22% increase, the population of a village is 1464.
> What was the population originally?
> The multiplier for a 22% increase is 1.22
> The calculation is: $1464 \div 1.22 = 1200$
> So the original population was 1200.

Top Tip!

The multiplier method is much simpler and is shown below.

Questions

Grade C

1 a £500 is invested in an account that pays 3.5% compound interest.
How much will be in the account after 6 years?

b A rabbit colony has a disease that reduces its population by 9% each year. If the original population was 2000, how many rabbits will there be after 4 years?

Grade B

2 a In a sale, a TV is reduced to £322. This is an 8% reduction from the original price. What was the original price?

b The average attendance at Barnsley football club increased by 12% between 2005 and 2006. In 2006 the average attendance was 15 680. What was the average attendance in 2005?

Ratio

Ratios

- A **ratio** is a way of comparing the sizes of two or more quantities.

- A colon (:) is used to show ratios.   3 : 4 and 6 : 20 are ratios.

- A quantity can be divided into **portions** that are in a **given ratio**.

- The process has three steps.
    - Step 1: **Add** the separate parts of the ratio
    - Step 2: **Divide** this number into the original quantity
    - Step 3: **Multiply** this answer by the original parts of the ratio

> Share £40 in the ratio 2 : 3.
>
> Add 2 and 3 to find the total number of parts: 5
>
> Divide 40 by 5 to find the value of each part: 8
>
> Multiply each term in the ratio by 8: $2 \times 8 = 16$, $3 \times 8 = 24$
>
> So £40 divided in the ratio 2 : 3 gives shares of £16 and £24.

- When one part of a ratio is known, it is possible to calculate other values. Use the given information to find a **unit value** and use the unit value to find the required information.

> When the cost of a meal was shared between two families in the ratio 3 : 5, the smaller share was £22.50. How much did the meal cost altogether?
>
> $\frac{3}{8}$ of the cost was £22.50, so $\frac{1}{8}$ was £7.50. The total cost was $8 \times 7.50 = £60$.

Speed, time and distance

- **Speed**, **time** and **distance** are connected by the formula: **distance = speed × time**

- This formula can be rearranged to give: $\textbf{speed} = \dfrac{\textbf{distance}}{\textbf{time}}$ $\textbf{time} = \dfrac{\textbf{distance}}{\textbf{speed}}$

- Problems involving speed actually mean **average speed**, as maintaining a constant speed is not possible over a journey.

> A car travels at 40 mph for 2 hours. How far does it travel in total?
>
> distance = speed × time = $40 \times 2 = 80$ miles

- Use this diagram to remember the formulae that connect speed, time and distance.

Questions

Grade C

1 Divide the following amounts in the given ratios.
- **a** £500 in the ratio 1 : 4
- **b** 300 grams in the ratio 1 : 5
- **c** £400 in the ratio 3 : 5
- **d** 240 kg in the ratio 1 : 2

Grade C

2 a A box of crisps has two flavours, plain and beef, in the ratio 3 : 4. There are 42 packets of plain crisps. How many packets of beef crisps are there in the box?

b The ratio of male teachers to female teachers in a school is 3 : 7. If there are 21 female teachers, how many teachers are there in total?

Grade D

3 a A motorist travels a distance of 75 miles in 2 hours. What is his average speed?

b A cyclist travels for $3\frac{1}{2}$ hours at an average speed of 15 km per hour.
How far has she travelled?

D

Direct proportion problems

- When solving **direct proportion** problems, work out the **cost of one item**. This is called the **unitary method**.

> If eight cans of cola cost £3.60, how much do five cans cost?
>
> The cost of one can is 360 ÷ 8 = 45p, so five cans cost 5 × 0.45 = £2.25.

Top Tip!
Always check that the answer makes sense when compared with the numbers in the original problem.

D

Best buys

- Many products are sold in different sizes at different prices.

- To find a **best buy**, work out how much of the item you get for a unit cost, such as how much you get per penny or how much per pound.

- Always divide the quantity by the cost.

> A 400 g jar of coffee costs £1.44. A kilogram jar of the same coffee costs £3.80. Which jar is better value?
>
> 400 ÷ 144 = 2.77 g/penny   1000 ÷ 380 = 2.63 g/penny
>
> Hence the 400 g jar is better value.

Top Tip!
Be careful with units. Change pounds into pence and kilograms into grams.

C

Density

- **Density** is the mass of a substance per unit volume.

- Density is usually expressed in grams per cubic centimetre (g/cm³) or kilograms per cubic metre (kg/m³).

- The relationship between density, mass and volume is:

$$\text{density} = \frac{\text{mass}}{\text{volume}}$$

- The following triangle can be used to help remember the formulae that connect mass, density and volume.

mass = density × volume
density = mass ÷ volume
volume = mass ÷ density

Top Tip!
Notice the use of mass rather than weight in this relationship. Although there is a difference between mass and weight, in GCSE mathematics they are assumed to have the same meaning.

Questions

Grade D

1 a 40 bricks weigh 50 kg. How much will 25 bricks weigh?

b How many bricks are there on a pallet weighing 200 kg, assuming the weight of the pallet is zero?

Grade D

2 a A large tube of toothpaste weighs 250 g and costs £1.80. A travel-size tube contains 75 g and costs 52p. Which is better value?

b Which is the better mark: 62 out of 80 or 95 out of 120?

Grade C

3 a Calculate the density of a brick with a mass of 2500 grams and a volume of 400 cm³. State the units in your answer.

b Calculate the weight, in kilograms, of a piece of metal 3000 cm³ in volume, if its density is 9 g/cm³.

c A piece of wood has an average density of 52 kg/m³. It weighs 15.6 kg.

Find its volume. Give your answer in cubic metres.

Powers and reciprocals

Square roots and cube roots

- The **square root** of a given number is a number which, when multiplied by itself, produces the given number.

  > The square root of 16 is 4, since $4 \times 4 = 16$.

- Every positive number has two square roots, one positive and one negative.

  > Because -4×-4 also equals 16, $\sqrt{16} = \pm 4$.

- A square root is represented by the **symbol** $\sqrt{}$.

  > $\sqrt{25} = 5$

- Calculators have a '**square root' button**.

- Taking a square root is the **inverse operation** to squaring.

- The **cube root** of a given number is a number which, when multiplied by itself twice, produces the given number.

  > The cube root of 125 is 5, since $5 \times 5 \times 5 = 125$.

- A **cube root** is represented by the symbol $\sqrt[3]{}$.

  > $\sqrt[3]{125} = 5$

- Many calculators have a '**cube root' button**.

D-C

Powers

- **Powers** are a convenient way of writing repetitive multiplications.

  > $3 \times 3 \times 3 \times 3 \times 3 \times 3 = 3^6$, which is read as '3 to the power 6'.

- Powers are also called **indices** (singular **index**).

- The **power '2'** has a special name: '**squared**'.

- The **power '3'** has a special name: '**cubed**'.

D-C

Negative indices

- A negative index is a convenient way of writing the **reciprocal** of a number or term.

  > $2^{-2} = \frac{1}{2^2} = \frac{1}{4}$ $\qquad$ $\frac{1}{27} = \frac{1}{3^3} = 3^{-3}$

B

Questions

Grades D-C

1 a Write down the value of each of these.

  i $\sqrt{81}$ **ii** $\sqrt[3]{64}$ **iii** $\sqrt{25}$

  b Find the values of x in each of these expressions.

  i $x^2 = 4$ **ii** $x^3 = 1$ **iii** $x^3 = -8$

  c Use a calculator to find the value of each of these.

  i $\sqrt{576}$ **ii** $\sqrt[3]{1.331}$ **iii** $\sqrt{37.21}$

Grades D-C

2 a Write down the value of each of these cubes.

  i 3^3 **ii** 4^3 **iii** 10^3

  b Write these expressions, using power notation.

  i $4 \times 4 \times 4 \times 4 \times 4$

  ii $6 \times 6 \times 6 \times 6 \times 6 \times 6$

  iii $10 \times 10 \times 10 \times 10$

  iv $2 \times 2 \times 2 \times 2 \times 2 \times 2 \times 2$

  c Use a calculator to work out the values of the power terms in part **b**.

  d Continue the powers of 2 sequence for 10 terms.

    2, 4, 8, 16, 32, … , … , … , … , …

Grade B

3 a Write down each of these in fraction form.

  i 4^{-3} **ii** 7^{-1} **iii** 3^{-2}

  b Write down each of these in negative index form.

  i $\frac{1}{8}$ **ii** $\frac{1}{3}$ **iii** $\frac{1}{x^n}$

Remember: You must revise all content from Grade E to the level that you are currently working at.

Standard form

B

- **Standard form** (also called standard index form) is a way of writing very large or very small numbers, using powers of 10.

- Every standard form number is of the form $a \times 10^n$, where $1 \leqslant a < 10$. and n is a positive or negative whole number.

$$45\,000 = 4.5 \times 10^4 \qquad 0.000\,089 = 8.9 \times 10^{-5}$$

- Standard form numbers can be combined, using the **rules of powers**.

Work out $3 \times 10^2 \times 5 \times 10^3$.
$3 \times 10^2 \times 5 \times 10^3 = 3 \times 5 \times 10^2 \times 10^3 = 15 \times 10^5 = 1.5 \times 10^6$

Work out $(4.5 \times 10^3) \div (5 \times 10^7)$.
$(4.5 \times 10^3) \div (5 \times 10^7) = 4.5 \div 5 \times 10^3 \div 10^7 = 0.9 \times 10^{-4} = 9 \times 10^{-5}$

Rational numbers

C

- A **rational number** is any number that can be expressed as a **fraction**.

- Some fractions result in **terminating decimals** and some fractions result in **recurring decimals**.

$\frac{1}{16} = 0.0625$ which is a terminating decimal.
$\frac{1}{3} = 0.3333...$ which is a recurring decimal.

- Recurrence is shown by a dot or dots over the recurring digit or digits.

$0.3333...$ becomes $0.\dot{3}$ $0.277\,777...$becomes $0.2\dot{7}$
$0.518\,518\,518...$ becomes $0.\dot{5}1\dot{8}$

Finding reciprocals

C

- The **reciprocal** of a number is the result of **dividing** the number **into 1**.

- The reciprocal of a **fraction** is simply the fraction turned **upside down**.

The reciprocal of 5 is $\frac{1}{5}$ or $1 \div 5 = 0.2$.

The reciprocal of $\frac{4}{7}$ is $\frac{7}{4} = 1\frac{3}{4}$.

Questions

Grades B-A

1 a Write these numbers in standard form.
  i 560 000   **ii** 0.000 007   **iii** 3 million
b Write each of these as an ordinary numbers.
  i 6.4×10^6 **ii** 8.3×10^{-4} **iii** 9×10^8
c Work out the following. Give your answers in standard form
  i $5.2 \times 10^4 \times 3 \times 10^2$
  ii $(3.6 \times 10^6) \div (9 \times 10^2)$
  iii $1.8 \times 10^2 \times 5 \times 10^4$
  iv $(2.4 \times 10^4) \div (3 \times 10^8)$

Grade C

2 Write each of the following using recurrence notation.
  a 0.363 636...   **b** 0.615 615 615...
  c 0.366 6...   **d** $\frac{2}{3}$
  e $\frac{1}{6}$ **f** $\frac{7}{9}$

Grade C

3 a Work out the reciprocal of each number. Give your answers as terminating or recurring decimals.
  i 4   **ii** 20   **iii** 9
b Write down the reciprocals of the each fraction. Give your answers as mixed numbers.
  i $\frac{7}{8}$ **ii** $\frac{5}{9}$ **iii** $\frac{3}{13}$

Remember: You must revise all content from Grade E to the level that you are currently working at.

Multiplying and dividing powers

- When we **multiply** together powers of the same number or variable, we **add** the indices.

$$2^2 \times 2^3 = 2^{(2+3)} = 2^5 = 32 \qquad x^2 \times x^6 = x^{(2+6)} = x^8$$

- When we **divide** two powers of the same number or variable, we **subtract** the indices.

$$3^5 \div 3^2 = 3^{(5-2)} = 3^3 = 27 \qquad x^9 \div x^6 = x^{(9-6)} = x^3$$

- When you are working with more complex expressions the same rules apply. Treat each letter and number separately.

$$\frac{4a^3b^3 \times 2ab^2}{2a^2b} = \frac{4 \times 2 \times a^3 \times a \times b^3 \times b^2}{2a^2b} = \frac{8a^4b^5}{2a^2b} = 4a^2b^4$$

Top Tip!

Don't get confused with ordinary numbers when powers are involved. The rules above only apply to powers, so $2x^3 \times 3x^2 = 2 \times 3 \times x^3 \times x^2 = 6x^5$ *not* $5x^5$.

C-B

Power of a power

- When we **raise** a power term to a further power, we **multiply** the powers.

$$(a^2)^3 = a^6 \qquad (2x^3)^4 = 2^4 \times (x^3)^4 = 16x^{12}$$

B-A

Top Tip!

A negative power does not mean the answer will be negative. It means take the reciprocal.

Indices of the form $\frac{1}{n}$ and $\frac{a}{b}$

- An index of the form $\frac{1}{n}$ is the **nth root** of that number.

$$49^{\frac{1}{2}} = \sqrt{49} = 7 \qquad 8^{\frac{1}{3}} = \sqrt[3]{8} = 2 \qquad 36^{-\frac{1}{2}} = \frac{1}{\sqrt{36}} = \frac{1}{6}$$

- An index of the form $\frac{a}{b}$ is the ath power of the bth root. $\quad 32^{\frac{4}{5}} = (32^{\frac{1}{5}})^4 = 2^4 = 16$

- If the power is **negative**, take the **reciprocal** of the answer.

- Solve these problems in three steps.

  – Step 1: Take the root of the base number given by the denominator of the fraction.

  – Step 2: Raise the result to the power given by the numerator of the fraction.

  – Step 3: Take the reciprocal (divide into 1) of the answer, which is what a negative power tells you to do.

$$125^{-\frac{2}{3}} = (125^{-\frac{1}{3}})^{-2} = (5)^{-2} = \frac{1}{25}$$

A

Questions

Grades C-B

1 a Write each of the following as a single power.

   i $2^3 \times 2^4$ **ii** $2^4 \times 2^5$ **iii** $x^6 \times x^3$

   iv $3^5 \div 3^2$ **v** $3^8 \div 3^4$ **vi** $x^7 \div x^3$

b Simplify these expressions.

   i $\dfrac{2a^2bc^3 \times 4ab^2c}{2abc}$ **ii** $\dfrac{6xyz \times 2x^2y^3z^2 \times 3xz}{9x^2yz^3}$

Grades B-A

2 Simplify these expressions.

   a $(x^3)^5$ **b** $(3a^2)^3$ **c** $(2xy^2)^2$

Grade A

3 Evaluate the following.

   a $25^{\frac{1}{2}}$ **b** $64^{\frac{1}{3}}$ **c** $125^{-\frac{1}{3}}$

Grade A

4 Evaluate the following.

   a $64^{-\frac{2}{3}}$ **b** $27^{-\frac{2}{3}}$ **c** $100^{-\frac{5}{2}}$

Surds

Surds

- **Surds** are numerical expressions that contain square roots. $\sqrt{2}$ and $\sqrt{5}$ are surds.

- Answers given as surds are **exact** answers. Answers written as $\sqrt{5} \approx 2.236$ are approximations.

- There are four rules that must be followed when working with surds.

  - $\sqrt{a} \times \sqrt{b} = \sqrt{ab}$
  - $\sqrt{a} \div \sqrt{b} = \sqrt{\frac{a}{b}}$
  - $C\sqrt{a} \times D\sqrt{b} = CD\sqrt{\frac{a}{b}}$
  - $C\sqrt{a} \div D\sqrt{b} = \frac{C}{D}\sqrt{\frac{a}{b}}$

  $\sqrt{2} \times \sqrt{6} = \sqrt{12}$ $\qquad$ $\sqrt{10} \div \sqrt{2} = \sqrt{5}$ $\qquad$ $3\sqrt{7} \times 2\sqrt{3} = 6\sqrt{21}$ $\qquad$ $8\sqrt{15} \div 2\sqrt{3} = 4\sqrt{5}$

- Expressions involving surds can usually be simplified.

  $\sqrt{20} = \sqrt{4 \times 5} = 2\sqrt{5}$ $\qquad$ $\sqrt{20} + \sqrt{45} = 2\sqrt{5} + 3\sqrt{5} = 5\sqrt{5}$

  $(2 + \sqrt{3})(2 - \sqrt{6}) = 2 \times 2 - 2\sqrt{6} + 2\sqrt{3} - \sqrt{18} = 4 - 2\sqrt{6} + 2\sqrt{3} - 3\sqrt{2}$

  $(4 + \sqrt{5})(4 - \sqrt{5}) = 16 - 4\sqrt{5} + 4\sqrt{5} - 5 = 11$

  $(2 + \sqrt{6})^2 = (2 + \sqrt{6})(2 + \sqrt{6}) = 4 + 2\sqrt{6} + 2\sqrt{6} + 6 = 10 + 4\sqrt{6}$

> **Top Tip!**
> Try to split numbers into a product that involves a square number.

Rationalising the denominator

- A denominator in surd form can be rationalised.

  Rationalise the denominator of $\frac{\sqrt{2}}{\sqrt{5}}$.

  Multiply the top and the bottom by $\sqrt{5}$: $\frac{2 \times \sqrt{5}}{\sqrt{5} \times \sqrt{5}} = \frac{2\sqrt{5}}{5}$

  Rationalise the denominator of $\frac{4\sqrt{2}}{\sqrt{6}}$.

  Multiply the top and the bottom by $\sqrt{6}$: $\frac{4\sqrt{2} \times \sqrt{6}}{\sqrt{6} \times \sqrt{6}} = \frac{4\sqrt{12}}{6} = \frac{4 \times 2\sqrt{3}}{6} = \frac{4\sqrt{3}}{3}$

> **Top Tip!**
> Multiply the top and bottom by the same surd that is in the denominator.

Solving problems with surds

- Surds can be used to solve problems.

  Show that this triangle is right-angled.

  If the triangle is right-angled then it will obey Pythagoras' theorem. (See pages 35–7.)

  $(\sqrt{2} + \sqrt{18})^2 + 6^2 = (\sqrt{2} + \sqrt{18})(\sqrt{2} + \sqrt{18}) + 36$
  $= 2 + \sqrt{36} + \sqrt{36} + 18 + 36 = 68$

  $\sqrt{68} = \sqrt{(4 \times 17)} = 2\sqrt{17}$. Hence it is right-angled.

Questions

1 Simplify each of the following. Leave your answers in surd form.

  a $\sqrt{12} \div \sqrt{3}$ $\qquad$ **b** $2\sqrt{18} \times 3\sqrt{2}$

  c $\sqrt{75}$ $\qquad$ **d** $\sqrt{32} + \sqrt{50}$

  e $\sqrt{3}(2 + \sqrt{3})$ $\qquad$ **f** $(2 + \sqrt{5})(2 - \sqrt{5})$

  g $(3 - \sqrt{5})^2$

2 Calculate the area of the rectangle opposite, giving the answer in surd form.

3 Rationalise the denominators and simplify if possible.

  a $\frac{3}{\sqrt{12}}$ $\qquad$ **b** $\frac{3\sqrt{2}}{\sqrt{8}}$ $\qquad$ **c** $\frac{3 - \sqrt{5}}{\sqrt{5}}$

$2 + \sqrt{3}$

$5 - \sqrt{3}$

Variation

Direct variation

- **Direct variation** means the same as **direct proportion**.

- There is direct variation between two variables when one variable is a simple multiple of the other.

- When two variables are proportional to each other, you can set up an equation with a constant of proportionality.

> Pay is directly proportional to time worked.
> Pay $\propto$ Time $\Rightarrow$ Pay $= k \times$ Time,
> where k is the constant of proportionality.

- When a power or root of one variable is directly proportional to another variable, the process is the same as for a linear direct variation.

> y is directly proportional to the square of x.
> $y \propto x^2 \Rightarrow y = kx^2$, where k is the constant of proportionality.

- In an examination question, some numerical information will be given to enable the constant of proportionality to be found.

> y is directly proportional to the square root of x.
> When $y = 6$, $x = 4$.
> $y \propto \sqrt{x} \Rightarrow y = k\sqrt{x}$. $6 = k\sqrt{4} \Rightarrow 6 = 2k \Rightarrow k = 3$.

- A question will usually give the value of one variable and ask for the value of the other variable.

> y is directly proportional to the cube of x.
> If $y = 20$ when $x = 2$, what is the value of y when $x = 4$?
> $y \propto x^3 \Rightarrow y = kx^3$. $20 = k \times 2^3 \Rightarrow 20 = 8k \Rightarrow k = 2.5$.
> When $x = 4$, $y = 2.5 \times 4^3 = 2.5 \times 64 = 160$.

B-A

Inverse variation

- There is inverse variation between two variables when one variable decreases as the other increases such that the product of the two variables is constant.

- This is represented mathematically by the **reciprocal**.

y is inversely proportional to x. $\quad y \propto \frac{1}{x} \Rightarrow y = \frac{k}{x}$.

- Inverse variation problems involving powers, roots and given numerical information are solved using the same method as for direct variation problems.

y is inversely proportional to the cube root of x.

If $y = 20$ when $x = 8$, what is the value of y when $x = 64$?

$y \propto \frac{1}{\sqrt[3]{x}} \Rightarrow y = \frac{k}{\sqrt[3]{x}} \Rightarrow 20 = \frac{k}{2} \Rightarrow k = 40$.

When $x = 64$, $y = \frac{40}{\sqrt[3]{64}} = \frac{40}{4} = 10$.

A

Top Tip!

In a GCSE examination, writing down the proportionality equation gets a mark.

Questions

Grade A

1 y is directly proportional to x^2. When $y = 5$, $x = 10$.
  a Write down the equation of proportionality.
  b Find the constant of proportionality.
  c Find the value of y when $x = 5$.
  d Find the value of x when $y = 20$.

Grade A

2 y is inversely proportional to $\sqrt{x}$. When $y = 6$, $x = 4$.
  a Write down the equation of proportionality.
  b Find the constant of proportionality.
  c Find the value of y when $x = 9$.
  d Find the value of x when $y = 12$.

Limits

C-A

Limits of accuracy

- All recorded measurements will have been rounded off to some degree of accuracy.

- This defines the possible true values before the rounding took place, and hence the **limits of accuracy**.

- The range of values between the limits of accuracy is called the **rounding error**.

The length of a piece of paper is 24.7 cm accurate to one decimal place.
What are the limits of accuracy? $24.65 \leqslant$ length < 24.75

The length of a motorway is 220 km to three significant figures.
What are the limits of accuracy? $219.5 \leqslant$ length < 220.5

- The examples above all relate to **continuous** data. If the data is **discrete**, the limits of accuracy are different.

The number of sweets in a jar is 100 to the nearest 10. What is the smallest and the greatest number of sweets that can be in the jar?

The least is 95 and the most is 104.

Top Tip!

The limits of accuracy are always given to one more degree of accuracy than the rounded value.

A

Calculating with limits of accuracy

- When we combine two or more linear values, the errors in the linear measures will be compounded producing a larger error in the calculated value.

- It is important to combine the values together in the correct way. The table below shows the combinations to give the maximum and minimum values for the four rules of arithmetic.

Operation	Maximum	Minimum
Addition ($a + b$)	$a_{max} + b_{max}$	$a_{min} + b_{min}$
Subtraction ($a - b$)	$a_{max} - b_{min}$	$a_{min} - b_{max}$
Multiplication ($a \times b$)	$a_{max} \times b_{max}$	$a_{min} \times b_{min}$
Division ($a \div b$)	$a_{max} \div b_{min}$	$a_{min} \div b_{max}$

Top Tip!

Answers often come out with several decimal places. Never round off to less than three significant figures.

The distance from Penistone to Glossop is 16.8 miles to the nearest tenth of a mile. The time for the journey calculated on a GPS unit is 25 minutes to the nearest minute. Between what limits is the expected average speed of the journey? Speed = distance $\div$ time

maximum speed = maximum distance $\div$ minimum time (hours) = $16.85 \div \frac{24.5}{60}$ = 41.27 mph

minimum speed = minimum distance $\div$ maximum time (hours) = $16.75 \div \frac{25.5}{60}$ = 39.41 mph

The limits are $39.41 \leqslant$ speed < 41.27.

Questions

1 Write down the lower and upper limits of each of these values, rounded to the accuracy stated.

a 7 m (1 significant figures)

b 34 kg (nearest kg)

c 320 cm (2 significant figures)

2 a A rectangle is measured as 8 cm by 12 cm, measured to the nearest cm. What are the limits of the area of the rectangle?

b A journey of 32 miles, measured to the nearest mile, takes 60 minutes to the nearest 10 minutes. What are the limits of the average speed of the journey?

Number checklist

I can...

- [ ] work out one quantity as a fraction of another
- [ ] solve problems using negative numbers
- [ ] multiply and divide by powers of 10
- [ ] multiply together numbers that are multiples of powers of 10
- [ ] round numbers to one significant figure
- [ ] estimate the answer to a calculation
- [ ] order lists of numbers containing decimals, fractions and percentages
- [ ] multiply and divide fractions
- [ ] calculate with speed, distance and time
- [ ] compare prices to find 'best buys'
- [ ] find the new value after a percentage increase or decrease
- [ ] find one quantity as a percentage of another

You are working at (Grade D) level.

- [ ] work out a reciprocal
- [ ] recognise and work out terminating and recurring decimals
- [ ] write a number as a product of prime factors
- [ ] find the HCF and LCM of pairs of numbers
- [ ] use the index laws to simplify calculations and expressions
- [ ] multiply and divide with negative numbers
- [ ] multiply and divide with mixed numbers
- [ ] round numbers to given numbers of significant figures
- [ ] find a percentage increase
- [ ] work out compound interest problems
- [ ] solve problems using ratio

You are working at (Grade C) level.

- [ ] work out the square roots of decimal numbers
- [ ] estimate answers using square roots of decimal numbers
- [ ] work out reverse percentage problems
- [ ] solve problems involving density
- [ ] write and calculate with numbers in standard form
- [ ] find limits of numbers given to the nearest unit

You are working a (Grade B) level.

- [ ] solve complex problems involving percentage increases and decreases
- [ ] use the rules of indices for fractional and negative indices
- [ ] convert recurring decimals to fractions
- [ ] simplify surds
- [ ] find limits of numbers given to various accuracies

You are working at (Grade A) level.

- [ ] solve problems using surds
- [ ] solve problems using combinations of numbers rounded to various limits

You are working at (Grade A*) level.

Circles and area

Grades

Circumference of a circle

D-C

- The **circumference** of a circle is the distance around the circle (the **perimeter**).

- The circumference of a circle is given by the formula:

$C = \pi d$ or $C = 2\pi r$

where d is the **diameter** and r is the **radius**.

> Calculate the circumference of the circle shown.
> Give your answer to 3 significant figures.
> $C = \pi d$
> $= \pi \times 5$ cm $= 15.7$ cm (to 3 sf)

5 cm

> **Top Tip!**
> You can learn just one formula because you can always find the radius from the diameter, or vice versa, as: $d = 2r$

- The value of π is a **decimal** that goes on forever but it is 3.142 to three decimal places.

- All calculators should have a π **button** which will give the value as 3.141592654.

- The formula you use depends on whether you are given the radius or the diameter.

- On the non-calculator paper, you may be asked to give an answer in terms of π. In this case leave your answer as, for example, 6π.

Area of a circle

D-C

- The area of a circle is the **space inside** the circle.

- The area of a circle is given by the formula: $A = \pi r^2$ where r is the radius.

- You must use the radius when calculating the area.

8 cm   4 cm

> **Top Tip!**
> Unless you are asked to give your answer in terms of π, use a calculator to work out circumferences and areas of circles and give answers correct to at least one decimal place.

Area of a trapezium

D-C

- To work out the area of a trapezium, multiply half the sum of the parallel sides by the distance between them using the formula:

area $= \frac{1}{2} \times (a + b) \times h$

Find the area of this trapezium.

$A = \frac{1}{2} \times (8 + 12) \times 6$
    $= \frac{1}{2} \times 20 \times 6$
    $= 60$ m^2

8 m

6 m

12 m

Questions

Grades D-C

1 a Calculate the circumference of a circle with a diameter of 12 cm.
Give your answer to 1 decimal place.

b Calculate the circumference of a circle with a radius of 4 cm.
Leave your answer in terms of π.

Grades D-C

2 a Calculate the area of a circle with a radius of 15 cm. Give your answer to 1 decimal place.

b Calculate the area of a circle with a radius of 3 cm.
Leave your answer in terms of π.

Grade D

3 Find the area of each of these trapezia.

a
5 cm
5 cm
10 cm

b
7 cm
5 cm
4 cm
10 cm

Sectors and prisms

Sectors

- A **sector** is part of a circle, bounded by two radii of the circle and one of the **arcs** formed by the intersection of these radii with the circumference.

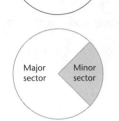

- When a circle is divided into only two sectors, the larger one is called the **major sector** and the smaller one is called the **minor sector**.

- The length of the arc of a sector is given by the formula: **arc length** $= \frac{\theta°}{360°} \times 2\pi r$ or $\frac{\theta°}{360°} \times \pi d$ where θ is the angle at the centre and r is the radius.

- The area of a sector is given by the formula: **sector area** $= \frac{\theta°}{360°} \times \pi r^2$

Top Tip!

On the non-calculator paper, you may be asked to give an answer in terms of π. In this case, the sector angle will be a factor of 360 such as 18, 20, or 30.

Find the arc length and the area of the sector in the diagram.

The sector angle is 35° and the radius is 6 cm.
Therefore,

arc length $= \frac{35°}{360°} \times \pi \times 2 \times 6 = 3.67$ cm

sector area $= \frac{35°}{360°} \times \pi \times 6^2 = 11.0$ cm^2

Prisms

C

- A **prism** is a 3-D shape that has the same cross-section all the way through it.

- The **surface area** of a prism is the area covered by its net.

- The **volume** of a prism is found by multiplying its cross-sectional area by the length of the prism:
  volume of a prism = area of cross-section × length or $V = Al$

Calculate the surface area and the volume of this triangular prism.
The surface area is made up of the area of the three rectangles and the area of the two isosceles triangles.
Area of the three rectangles $= 10 \times 5 + 10 \times 5 + 10 \times 6$
$= 50 + 50 + 60 = 160$ cm^2
Area of one triangle $= \frac{6}{2} \times 4 = 12$ cm^2, so area of two triangles $= 24$ cm^2
Therefore, the total surface area is 184 cm^2.
Area of the cross-section = area of one triangle = 12 cm^2.
$V = Al$. So, $V = 12 \times 10 = 120$ cm^2

Questions

Grade A-A*

1 a Calculate the arc length and area of this sector.

b A sector of a circle, radius 6 cm, has an angle of 40°.

  i Calculate the perimeter of the sector.

  ii Calculate the area of the sector.

  Give your answers in terms of π.

Grade C

2 a A cuboid has dimensions 4 cm by 6 cm by 10 cm.

  i Calculate the area of its cross-section.

  ii Calculate its volume.

b A triangular prism has a cross-sectional area of 3.5 m^2 and a length of 12 m. Calculate its volume.

Cylinders and pyramids

C

Volume of a cylinder

- A **cylinder** is a **prism** with a **circular cross-section**.

- The formula for the **volume of a cylinder** is:

  $V = \pi r^2 h$

  where r is the radius and h is the height or length of the cylinder.

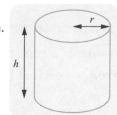

Top Tip!

You should learn the formula $V = \pi r^2 h$. It is not given on the formula sheet.

A

Surface area of a cylinder

- The **surface area** of a cylinder is made up of its **curved surface area** plus the area of its two **circular ends**.

- The curved surface opens out to a rectangle with length equal to the circumference of the circular end.

- The two circular ends each have an area of πr^2.

- The formula for the total surface area is:

  total surface area = $2\pi rh + 2\pi r^2$ or $\pi dh + 2\pi r^2$

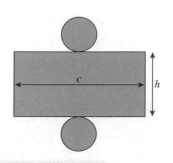

  What is the total surface area of a cylinder with a radius of 5 cm and a height of 12 cm?

  Total surface area = $2 \times \pi \times 5 \times 12 + 2 \times \pi \times 5^2 = 534.1$ cm^2

A

Volume of a pyramid

- A **pyramid** is a 3-D shape with a base from which triangular faces rise to a common **vertex**.

- The base can be any **polygon**, but is usually a **triangle**, **square** or **rectangle**.

- The **volume** of a pyramid is given by the formula:

  V = $\frac{1}{3}Ah$

  where A is the base area and h is the vertical height.

- A triangular-based pyramid is called a **tetrahedron**.

  Calculate the volume of the pyramid on the right.

  Base area = $6 \times 8 = 48$ cm^2

  Volume = $\frac{1}{3} \times 48 \times 10 = 160$ cm^3

Questions

Grade C

1 Calculate the volume of a cylinder with a radius of 4 cm and a height of 10 cm. Give your answer in terms of π.

Grade A

2 Calculate the surface area of the cylinder in question **1**. Give your answer in terms of π.

Grade A

3 Calculate the volume of a pyramid with a square base of side 11 cm and a vertical height of 15 cm.

Cones and spheres

Cones

- A cone can be treated as a pyramid with a **circular base**.

- The formula for the **volume** of a cone is the same as that for a pyramid:
  Volume = $\frac{1}{3}$ × base area × vertical height
  $V = \frac{1}{3}\pi r^2 h$

  where r is the radius of the base and h is the vertical height of the cone.

- The **curved surface area** of a cone is given by:

  $S = \pi r l$

  where l is the **slant height** of the cone.

- The **total surface area** of a cone consists of the curved surface area plus the area of its circular base.

- The **total surface area** is given by:

  $A = \pi r l + \pi r^2$

> **Top Tip!**
> The formula for the volume and curved surface area of a cone are given on the formula sheet that is included with the examination. However, it is much better to learn them.

Calculate the volume and the total surface area of the cone shown.
Volume $\frac{1}{3} \times \pi \times 5^2 \times 12 = 314.2$ cm³
Total surface area $= \pi \times 5 \times 13 + \pi \times 5^2 = 282.7$ cm²

13 cm • 12 cm • 10 cm

Spheres

- A sphere is a **ball** shape.

- The Earth and other planets are approximately spherical.

- The **volume** of a **sphere** is given by the formula:
  $V = \frac{4}{3}\pi r^3$

- The **surface area** of a sphere is given by the formula:
  $A = 4\pi r^2$

Calculate the volume and the surface area of a sphere with a radius of 9 cm.
Volume $= \frac{4}{3} \times \pi \times 9^3 = 3053.6$ cm³
Surface area $= 4 \times \pi \times 9^2 = 1017.9$ cm²

Questions

Grade A

1 For a cone with a radius of 6 cm, a vertical height of 8 cm and a slant height of 10 cm, calculate
  i its volume and   **ii** its curved surface area.

Grade A

2 For a sphere with a radius of 3 cm, calculate
  i its volume and   **ii** its surface area.
  Give your answers in terms of π.

Grade A

3 For a hemisphere with a radius of 5 cm, calculate
  i its volume and   **ii** its **total** surface area.

Pythagoras' theorem

Pythagoras' theorem

C

- **Pythagoras' theorem** connects the sides of a **right-angled triangle**.

- Pythagoras' theorem states that:

  'In any right-angled triangle, the square of the hypotenuse is equal to the sum of the squares of the other two sides.'

- The **hypotenuse** is the **longest** side of the triangle, which is always **opposite** the right angle.

- Pythagoras' theorem is usually expressed as a formula: $c^2 = a^2 + b^2$

- The formula can be rearranged to find one of the other sides, for example: $a^2 = c^2 - b^2$

Top Tip!
To find the actual value of a side, don't forget to take the square root.

Finding lengths of sides

C

- If you **know the lengths of two sides** of a right-angled triangle, you can always Use Pythagoras' theorem to **find the length of the third side**.

Find the length of the hypotenuse of this triangle.

Using Pythagoras' theorem
$x^2 = 9^2 + 6.2^2$
$= 81 + 38.44$
$= 119.44$
$x = \sqrt{119.44}$
$= 10.9$ cm

Find the length of the side marked y in this triangle.

Using Pythagoras' theorem
$y^2 = 16^2 - 12^2$
$= 256 - 144$
$= 112$
$y = \sqrt{112}$
$= 10.6$ cm

Real-life problems

C

- Pythagoras' theorem can be used to solve practical problems.

A ladder of length 5 m is placed with the foot 2.2 m from the base of a wall. How high up the wall does the ladder reach?
$x^2 = 5^2 - 2.2^2$
$= 25 - 4.84$
$= 20.16$
$x = \sqrt{20.16} = 4.49$ m

Questions

Grade C

1 Explain why a triangle with sides of 3 cm, 4 cm and 5 cm is right-angled.

Grade C

2 a Work out the length of the diagonal of a rectangle with sides 10 cm and 5 cm.

b A ladder is placed 1.5 m from the base of the wall and reaches 3.6 m up the wall. How long is the ladder?

Pythagoras' theorem and isosceles triangles

- Every **isosceles triangle** has a line of symmetry that divides it into **two congruent** right-angled triangles.

Find the area of this triangle.

Split the triangle into two right-angled triangles with sides of 3.5 cm and 9 cm.

Use Pythagoras' theorem to find the height (h): $h = \sqrt{(9^2 - 3.5^2)} = 8.29$ cm

Area $= \frac{1}{2} \times 7 \times 8.29 = 29.0$ cm^2

9 cm   9 cm

7 cm

Pythagoras' theorem in three dimensions

- 3-D problems in GCSE examinations can be solved using right-angled triangles and trigonometry (see page 37).

- Solve these problems in four steps.
  - Step 1: Identify the right-angled triangle that includes the required information
  - Step 2: Redraw and label this triangle with the given lengths and the length to be found
  - Step 3: Use your diagram to decide whether it is the hypotenuse or one of the shorter sides which has to be found
  - Step 4: Solve the problem and round to a suitable degree of accuracy

Top Tip!

The length of a diagonal in any cuboid with sides a, b and c is given by:
$\sqrt{(a^2 + b^2 + c^2)}$

Find the distance BH in this cuboid.

First identify a right-angled triangle that contains the side BH and draw it.

This gives triangle BEH, which contains two lengths that you do not know, EB and BH. Let EB = x and BH = y.

Next identify a right-angled triangle that contains the side EB and draw it.

This gives triangle EAB. You can now find EB.

By Pythagoras' theorem

$x^2 = 15^2 + 25^2 = 850$ (there is no need to find x)

Use triangle BEH to find BH.

By Pythagoras' theorem

$y^2 = 10^2 + 850 = 950$

So y = BH = $\sqrt{950}$ = 30.8 cm

Questions

Grade B

1 Find the area of an isosceles triangle with sides of 10 cm, 10 cm and 8 cm.

Grade A

2 Find the length of a diagonal in a cuboid with sides of 3 cm, 4 cm and 5 cm.

Grades

B

A

Using Pythagoras and trigonometry

Trigonometry

B

- **Trigonometry** is concerned with the calculation of sides and angles in triangles.
- There are three trigonometric ratios: **sine**, **cosine** and **tangent**. These are abbreviated to sin, cos and tan.
- The sine, cosine and tangent ratios are defined in terms of the sides of a right-angled triangle and an angle, θ:
- Calculators have **sine**, **cosine** and **tangent buttons**.
- To calculate angles, use the **inverse functions** on your calculator, usually marked $\sin^{-1}$, etc. $\sin 45° = 0.707$ $\cos^{-1} 0.4 = 66.4°$ $\tan^{-1} 0.5 = 26.6°$

Hypotenuse (H)

Opposite (O)

θ

Adjacent (A)

Sine

B

- There are three possible rearrangements of the definition of sine:

$\sin \theta = \dfrac{opp}{hyp}$ $opp = hyp \times \sin \theta$ $hyp = \dfrac{opp}{\sin \theta}$

Use the sine ratio to find the value of x in this triangle.

x is opposite the known angle, so use $opp = hyp \times \sin \theta$

$x = 6 \times \sin 35° = 3.44$ cm

6 cm

x

35°

Top Tip!

All calculators work differently. Make sure you can use yours to get the answer shown.

Cosine

B

- There are three possible rearrangements of the definition of cosine:

$\cos \theta = \dfrac{adj}{hyp}$ $adj = hyp \times \cos \theta$ $hyp = \dfrac{adj}{\cos \theta}$

Use the cosine ratio to find the value of x in this triangle.

x is the hypotenuse, so use $hyp = \dfrac{adj}{\cos \theta}$

$x = 8 \div \cos 42° = 10.8$ cm

x

42°

8 cm

Top Tip!

Make sure your calculator is set to degrees. You will lose marks for using other angle measures.

Tangent

B

Top Tip!

The definitions of sine, cosine and tangent are not given in the formula sheets, so you need to learn them. A lot of people use SOHCAHTOA as a mnemonic.

- There are three possible rearrangements of the definition of tangent:

$\tan \theta = \dfrac{opp}{adj}$ $opp = adj \times \tan \theta$ $adj = \dfrac{opp}{\tan \theta}$

Use the tangent ratio to find the angle x in this triangle.

x is the angle, so use $\tan \theta = \dfrac{opp}{adj}$

$\tan x = 7 \div 10 = 0.7$, so $x = \tan^{-1} 0.7 = 35.0°$

7 cm

x

10 cm

Questions

Grade B

1 a Use your calculator to write down, to 3 significant figures, the value of:

   i $\sin 31°$ **ii** $\cos 55°$ **iii** $\tan 10°$

b Use your calculator to write down, to 1 decimal place, the value of:

   i $\sin^{-1} 0.6$ **ii** $\cos^{-1} 0.7$ **iii** $\tan^{-1} 2$

Grade B

2 Use the appropriate trigonometric ratio to find the values marked x in the triangles below.

8 cm

16 cm

x

12 cm

46°

x

4 cm

28°

x

Grades

SHAPE, SPACE AND MEASURES

Which ratio to use

- The difficulty with any trigonometric problem is knowing which ratio to use to solve it.

- If the problem involves **two sides and no angles**, use **Pythagoras' theorem** to find the missing side.

- If the problem has **an angle** in it, then you will need to use **trigonometry**.

- There are four steps when solving a trigonometry problem.
  - Step 1: Identify what information is given and what information needs to be found
  - Step 2: Decide which ratio to use
  - Step 3: Set up the appropriate calculation
  - Step 4: Work out the answer and round to an appropriate degree of accuracy

Top Tip!
Know your calculator. On some calculators it is better to type in 7 × cos 30 and on others cos 30 × 7.

Find the length of the side marked x in this triangle.

Step 1: Identify what information is given and what needs to be found. Namely, x is adjacent to the angle and 7 cm is the hypotenuse.

Step 2: Decide which ratio to use. Only one ratio uses adjacent and hypotenuse: **cosine**.

Step 3: Set up the calculation. Remember adj = hyp × cos θ, so the calculation is $x = 7 \times \cos 30°$.

Step 4: Work it out. $x = 6.06$ cm.

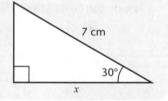

Solving problems

- Trigonometric problems in GCSE are usually set in real-life situations.

- The key is to find a right-angled triangle and then solve it as described above.

Top Tip!
Always redraw and label the triangle. This avoids confusion with any other numbers that might be on the diagram provided.

For health and safety reasons a ladder should be placed at 15° to the vertical. A ladder is 4 m long. How far from a wall should the ladder be placed?

Step 1: Draw the triangle and identify the relevant information. h is opposite to the angle and 4 m is the hypotenuse.

Step 2: Decide which ratio to use. Only one ratio uses opposite and hypotenuse: **sine**.

Step 3: Set up the calculation. Remember opp = hyp × sin θ, so the calculation is $h = 4 \times \sin 15°$.

Step 4: Work it out. $h = 1.04$ metres.

Questions

Grade B
1 Find the angle or length marked x in each of the triangles below.

Grade B
2 The angle of elevation to a church spire is measured as 18° from a point 80 m from the base. Calculate the height of the spire.

Geometry

Special quadrilaterals

D-C

- There are many different quadrilaterals. You will already know the **square** and the **rectangle**.

- The **square** has the following properties:
  – all sides are equal
  – all angles are equal
  – diagonals bisect each other and cross at right angles

Square

- The **rectangle** has the following properties:
  – opposite sides are equal
  – all angles are equal
  – diagonals bisect each other

Rectangle

- The **parallelogram** has the following properties:
  – opposite sides are equal and parallel
  – opposite angles are equal
  – diagonals bisect each other

Parallelogram

- The **rhombus** has the following properties:
  – all sides are equal
  – opposite sides are parallel
  – opposite angles equal
  – diagonals bisect each other and cross at right angles

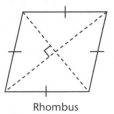

Rhombus

- The **kite** has the following properties:
  – two pairs of sides are equal
  – one pair of opposite angles are equal
  – diagonals cross at right angles

Kite

- The **trapezium** has the following properties:
  – two unequal sides are parallel
  – interior angles add to 180°
  – an **isosceles trapezium** has a **line of symmetry**

Trapezium

Regular polygons

C

- A **regular polygon** is a polygon with all its sides the same length.
  Here are three regular polygons.

- In any regular polygon, all of the **interior angles**, i, are equal, and all of the **exterior angles**, e, are equal.

- To find the interior angle of any regular polygon, **divide the angle sum** (S) (given by $S = 180(n – 2)°$) by the number of **sides** (n).

- To find the exterior angle of any regular polygon, **divide 360°** by the number of **sides**.

Questions

1 **a** Marcie says, 'All squares are rectangles.' Is she correct?

  b Milly says, 'All rhombuses are parallelograms.' Is she correct?

  c Molly says, 'All kites are rhombuses.' Is she correct?

  Give a reason for each answer.

2 What is **i** the interior angle and **ii** the exterior angle of:

  a a regular octagon

  b a regular nonagon

3 What is the connection between the interior and exterior angles of any regular polygon?

Circle theorems

Circle theorems

- Here are four circle theorems that you will need to know for your GCSE examination.

  – Angles at the circumference in the same segment of a circle are equal

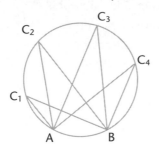

  – Every angle at the circumference of a semicircle that is subtended by the diameter of the semicircle is a right angle

  – The angle at the centre of a circle is twice the angle at the circumference subtended by the same arc

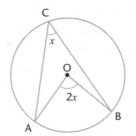

  – The sum of the opposite angles of a cyclic quadrilateral is 180°. (A cyclic quadrilateral is one in which all four verticies lie of the circumference of a circle.)

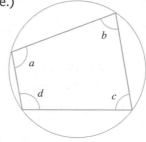

- These theorems can be used to find angles in circle problems.

O is the centre of each circle. Find the angles marked with letters in each of these diagrams. Give a reason for each answer.

  – w is 54° (angles in a semicircle, sum of angles in a triangle)
  – x is 43° (angle at the centre = twice the angle at circumference)
  – y is 28° (angles in the same segment are equal)
  – z is 98° (angles in a cyclic quadrilateral)

Questions

Grade B

1 O is the centre of the circle.
  a Find the angle marked *a*.
     Give a reason for your answer.
  b Find the angle marked *b*.
     Give a reason for your answer.

Grade B

2 O is the centre of the circle.
  a Find the angle marked *p*.
     Give a reason for your answer.
  b Find the angle marked *q*.
     Give a reason for your answer.

B-A

Tangents and chords

* A **tangent** is a straight line that touches a circle at only one point.

* A **chord** is a line that joins two points on the circumference of a circle.

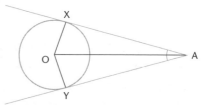

* There are some facts associated with tangents and chords that are useful in solving circle problems.

  – A tangent to a circle is perpendicular to the **radius** drawn to the point of contact

  – The line joining an external point to the centre of a circle bisects the angle between the tangents

  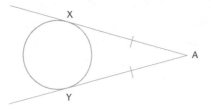

  – Tangents to a circle from an external point to the points of contact are equal in length

  – A radius perpendicular to a chord bisects the chord at 90° and when each end of the chord is joined to the centre of the circle, an isosceles triangle is formed.

* These facts can be used to find angles in circle problems.

O is the centre of both circles. Find the angles marked with letters in both of these diagrams. Give reasons for your answers.

x is 45° (radius is perpendicular to a tangent so there are two 90° angles in the quadrilateral hence 90° + 90° + 3x + x = 360°)

y is 124° (the third angle in an isosceles triangle in which the other two angles are 28°)

A

Alternate segment theorem

* The alternate segment theorem states that the angle between a tangent and a chord through the point of contact is equal to the angle in the alternate segment.

Find the angle marked x in the diagram.

The angle in the alternate segment is 70°.

Therefore $x = 180 - 75 - 70 = 35°$.

Questions

Grade A

1 What is the value of x.
Give a reason for your answer.

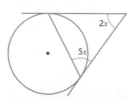

Grade A

2 What is the value of p.
Give a reason for your answer.

Transformation geometry

Transformations

- When a shape is **moved**, **rotated**, **reflected** or **enlarged**, it is called a **transformation**.
- There are four transfomations used in GCSE examinations: **translation**, **reflection**, **rotation** and **enlargement**.
- Shapes that are translated, reflected or rotated are **congruent** to the original shape.
- Shapes that are enlarged are **similar** to each other.
- The original shape is called the **object**, and the transformed shape is called the **image**.
- Sometimes the transformation is said to **map** the object onto the image.

— D-C

Congruent triangles

- Triangles that are exactly the same shape and size are called **congruent** triangles.
- A triangle must meet one or more of the **four conditions** below to be congruent.
  - **SSS** (side, side, side) means all three sides in one triangle are equal to the corresponding sides of the other triangle.
  - **SAS** (side, angle, side) means that two sides and the angle between them in one triangle are equal to the the two corresponding sides and the angle between them in the other triangle.
  - **ASA** (angle, side, angle) means that two angles and a side in one triangle are equal to the corresponding angles and side in the other triangle.
  - **RHS** (right angle, hypotenuse, side) means that both triangles have a right angle, their hypotenuse and another equal side are equal.

— B

Top Tip!

To check or work out a translation, use tracing paper. Trace the shape and then count squares as you move it horizontally and vertically.

Translations

- When a shape is **translated** it is moved without altering its orientation – it is not rotated or reflected.
- A **translation** is described by a vector. $\binom{-4}{5}$ is a vector.
- The **top number** in the vector is the movement in the x-direction. Positive values move to the right. Negative values move to the left.
- The **bottom number** in the vector is the movement in the y-direction. Positive values move upwards. Negative values move downwards.

Triangle C is translated from triangle A by the vector $\binom{0}{4}$.
Triangle B is translated to triangle D by the vector $\binom{-2}{4}$.

— D

Questions

(Grade B)

1 a State which, if any, of triangles A to D below are congruent.

b State which, if any, of triangles A to D are similar.

(Grade D)

2 Refer to the diagram above. What vector translates:

- **a** triangle A to triangle B
- **b** triangle A to triangle D
- **c** triangle C to triangle B
- **d** triangle C to triangle A

42

Remember: You must revise all content from Grade E to the level that you are currently working at.

Enlargements

D-C

- When a shape is **enlarged**, it changes its size to become a shape that is **similar** to the first shape.

- An **enlargement** is described by a **centre of enlargement** and a **scale factor**.

- The lengths of the sides of the image will be the lengths of the sides of the object **multiplied** by the **scale factor**.

- The centre of enlargement can be found by the **ray method**.

Top Tip!

Scale factors can also be fractions. In this case the image is smaller than the original object.

Triangle Q is an enlargement of triangle P with a scale factor of 3 from the centre O.

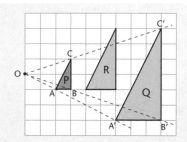

Combined transformations

B

- Transformations can be combined.

Triangle A is a rotation of the shaded triangle 90° clockwise about (0, 1).

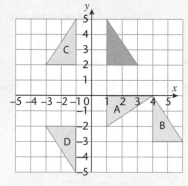

Triangle B is a rotation of triangle A 90° anticlockwise about (4, 0).

To get back to the shaded triangle, triangle B is translated by the vector $\binom{-3}{5}$.

Questions

Grade D

1 Refer to the diagram in the first example box.

  a i Write down the scale factor of the enlargement that takes triangle P to triangle R.

   ii Mark the centre of the enlargement that takes triangle P to triangle R.

  b i Write down the scale factor of the enlargement that takes triangle Q to triangle R.

   ii Mark the centre of the enlargement that takes triangle Q to triangle R.

Grade B

2 Refer to the diagram in the second example box.

  a Describe fully the transformation that takes the shaded triangle to triangle C.

  b Describe fully the transformation that takes triangle C to triangle D.

  c What **single** transformation takes triangle D to the shaded triangle?

Constructions

Constructing triangles

- When you are asked to **construct** a triangle, you are expected to use a ruler with a pair of **compasses** to measure lengths and a **protractor** to measure angles.

- There are **three ways** of constructing triangles.

- **All three sides given**
  - Use a ruler to draw one side. (Sometimes this side is already drawn.)
  - Use the ruler to set the compasses to the length of each of the other two sides, in turn, and draw arcs from the ends of the side you have drawn.
  - Join up the ends of the line to the point where the arcs intersect.

- **Two sides and the included angle given**
  - Use a ruler to draw one side. (Sometimes this side is already drawn.)
  - Use a protractor to measure and draw the angle at one end of the line you have drawn.
  - Set the compasses to the length of th other given side and draw an arc, cutting the line you have just drawn for the angle.
  - Join up the points where the arc cuts the line for the angle to the other end of the base line.

- **Two angles and the side between them given**
  - Use a ruler to draw the side. (Sometimes this side is already drawn.)
  - Use a protractor to measure and draw the angles at each end of the line.
  - Extend the lines to form the triangle.

Top Tip!

Always show your construction lines and arcs clearly. You will not get any marks if they can't be seen.

Constructing an angle of 60°

- You can use **compasses** and a **ruler** to draw an angle of 60° accurately.

  - Draw a line and mark a point where the angle will be drawn.

  - With the compasses centred on this point, draw an arc that cuts the line.

  - With the compasses set to the same radius, and centred on the point where the first arc cuts the line, draw another arc to cut the first arc.

  - Join the original point to the point where the arcs cross.

Questions

Use a ruler, compasses and protractor for these questions.

Grade D

1 a Draw accurately the triangle with sides of 4 cm, 5 cm and 6 cm, shown above.

b Draw accurately the triangle with sides of 6 cm and 5 cm and an included angle of 55°, shown above.

c Draw accurately the triangle with a side of 7 cm, and angles of 40° and 65° shown above.

Grade C

2 Follow the steps shown above and construct an angle of 60°.
Use a protractor to check the accuracy of your drawing.

(**Remember:**) You must revise all content from Grade E to the level that you are currently working at.

The perpendicular bisector

C

- To **bisect** means to **divide in half**.

- **Perpendicular** means **at right angles**.

- A **perpendicular bisector** divides a line in two, and is at right angles to it.
  - Start off with a line or two points. (Usually, these are given.)
  - Open the compasses to a radius about three-quarters of the length of the line, or of the distance between the points.
  - With the compasses centred on each end of the line (or each point) in turn, draw arcs on both sides of the line.
  - Join up the points where the arcs cross.

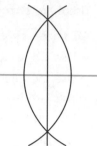

The angle bisector

C

- An **angle bisector** divides an angle into two smaller, equal angles.
  - Start off with an angle. (Usually, this is given.)
  - Open the compasses to any radius shorter than the arms of the angle. With the compasses centred on the vertex of the angle (the point where the arms meet) draw an arc on both arms of the angle.
  - Now, with the compasses still set to the same radius and centred on the points where these arcs cross the arms, draw intersecting arcs.
  - Join up the vertex of the angle and the point where the arcs cross.

The perpendicular at a point on a line

C

- The perpendicular at a point on a line is a line at right angles to the line, passing through the point.
  - Start off with a point on a line. (Usually, this is given.)
  - Open the compasses to a radius of about 3 cm and centred on the point, draw arcs on either side.
  - Now increase the radius of the compasses to about 5 cm and draw arcs centred on the two points on either side, so that they intersect, cutting the line.
  - Join the original point on the line and the point where the arcs cross.

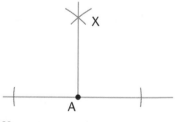

Questions

(Grade C)
1 Draw a line 6 cm long.
Following the steps above, draw the perpendicular bisector of the line.
Check that each side is 3 cm long and that the angle is 90°.

(Grade C)
2 Draw an angle of 70°.
Following the steps above, draw the angle bisector. Check that each half-angle is 35°.

(Grade C)
3 Draw a line and mark a point on it.
Following the steps above, draw the perpendicular at the point to the line. Check that the angle is 90°.

Constructions and loci

The perpendicular from a point to a line

- The **perpendicular** is a line that is at right angles to the original line and passes through the point.

  - Start with a line and a point not on the line. (Usually, these are given.)

  - Open the compasses to about 3 cm more than the distance from the point to the line. With the compasses centred on the point, draw arcs on the line on either side of the point.

  - Centring the compasses on the points where these arcs cut the line, draw arcs on the other side of the line so they intersect.

  - Join the original point and the point where the arcs cross.

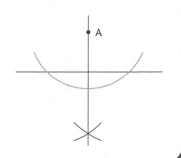

C

Loci

- A **locus** (plural **loci**) is the path followed by a point according to a rule.

  A point that moves so that it is always 4 cm from a fixed point is a circle of radius 4 cm.

  A point that moves so that it is always the same distance from two fixed points is the perpendicular bisector of the two points.

C

Practical problems

- **Loci** can be used to solve real-life problems.

  A horse is tethered to a rope 10 m long in a large, flat field. What is the area the horse can graze?

  The horse will be able to graze anywhere within a circle of radius 10 m.

- In reality the horse may not be able to graze an exact circle but the situation is modelled by the mathematics.

C

Questions

Grade C

1 Following the steps above, draw the perpendicular from a point to a line.
Check that the angle is 90°.

Grade C

2 A radar station in Edinburgh has a range of 200 miles.
A radar station in London has a range of 250 miles.
London and Edinburgh are 400 miles apart.
Sketch the area that the radar stations can cover.
Use a scale of 1 cm = 200 miles.

Similarity

B

Similar triangles

- Two triangles are **similar** if their corresponding angles are equal.
- The corresponding sides of similar triangles are in the same ratio.
- The ratio between the lengths of corresponding sides is called the scale factor.

Triangles ABC and PQR are similar.

a What is the scale factor between the triangles.

b Find the length PR.

a Comparing sides AB and PQ we can see that the scale factor is $10 \div 8 = 1.25$

b PR is, therefore, $10 \times 1.25 = 12.5$ cm

Special cases of similar triangles

- Some special cases of similar triangles frequently appear in GCSE examinations.
- The first is when two triangles share a common angle and parts of common sides.
- The key to solving this type of problem is to redraw the two triangles separately.

B

Find the sides marked x and y in these triangles.

Redraw the triangles.

Triangles AED and ABC are similar. So using the corresponding sides CB, DE with AC, AD $\frac{x}{10} = \frac{15}{6} \Rightarrow x = 10 \times \frac{15}{6} = 25$

Using the corresponding sides AB, AE with AC, AD gives

$\frac{8}{8+y} = \frac{15}{6} \Rightarrow 8 + y = 8 \times \frac{15}{6} = 20 \Rightarrow y = 12$

Questions

Grade B

1 To work out the height of a church, Zoe stands a 2 metre stick on the ground then lines up a point on the ground with the top of the stick and the top of the spire. She then paces out the distances.

Use similar triangles to work out the height of the spire.

Remember: You must revise all content from Grade E to the level that you are currently working at.

A

Similar shapes

- When two shapes are similar, there are relationships between their lengths, areas and volumes.

- The corresponding **lengths** of similar shapes are in the same ratio, the **length ratio**, which is equal to the **linear scale factor**.

- The **areas** of similar shapes are in the same ratio, called the **area ratio** or **area scale factor**. This is equal to the square of the length ratio.

- The **volumes** of similar 3-D shapes are also in the same ratio, called the **volume ratio** or **volume scale factor**. This is equal to the cube of the length ratio.

- If the linear scale factor is n, the area scale factor is n^2 and the volume scale factor is n^3.

> **Top Tip!**
> The key word in identifying these problems is **similar**, followed by a reference to either area or volume.

Every length in cuboid B is twice the corresponding length in cuboid A.

The length of A is 5 cm and the length of B is 10 cm, so the lengths are in the ratio 5 : 10 = 1 : 2.

The area of the front face of A is 15 cm² and the area of the front face of B is 60 cm², so the ratio of the areas is $15 : 60 = 1 : 4 = 1 : 2^2$.

The volume of cuboid A is 30 cm³ and the volume of cuboid B is 240 cm³, so the ratio of the volumes is $30 : 240 = 1 : 8 = 1 : 2^3$.

A*

Solving problems with area and volume ratios

- It is essential to follow three steps when solving problems about similar shapes
  - Step 1: Establish the given scale factor
  - Step 2: Establish the scale factor(s) required to solve the problem
  - Step 3: Multiply or divide the given dimension by the appropriate scale factor to find the required dimension

> **Top Tip!**
> If you are given the volume or area scale factor and you need to find the other of these, find the linear scale factor first as this makes working easier.

Soup is sold in similar small and large cans. The volume of the large can is 500 ml and the volume of the small can 300 ml.

The labels on the cans are also similar in shape. The label on the large can has an area of 40 cm². What is the area of the label on the small can?

Step 1: The volume scale factor is $500 \div 300 = 1.\dot{6}$

Step 2: The linear scale factor $= \sqrt[3]{1.\dot{6}}\ldots$ and the area scale factor $= (\sqrt[3]{1.\dot{6}})^2$

Step 3: Let the area of the label of the small can be A, then $A = 40 \div (\sqrt[3]{1.\dot{6}})^2 = 28.45$ cm².

Questions

Grade A

1 A cuboid is 3 cm by 8 cm by 12 cm. A similar cuboid is 9 cm by 24 cm by 36 cm.
  a What is the linear scale factor?
  b What is the area scale factor?
  c What is the volume scale factor?

Grade A*

2 Two similar statuettes are made from the same material. They are 15 cm and 25 cm tall respectively. The smaller statuette has a mass if 4 kg. If mass is proportional to volume, calculate the mass of the larger statuette.

Dimensional analysis

Dimensional analysis

- The **dimensions** of a shape include its length, width and height, and the products of these quantities.

- By looking at **formulae** it is possible to tell if they represent lengths, areas or volumes.

- All formulae that represent **lengths** are **one-dimensional**.

- All formulae that represent **lengths** have **single letters** in them.

- All formulae that represent **areas** are **two-dimensional**.

- All formulae that represent **areas** have **letters in 'pairs'** in them. A term such as r^2 means $r \times r$ so it is a 'pair' of letters.

- All formulae that represent **volumes** are **three-dimensional**.

- All formulae that represent **volumes** have **letters in 'threes'** in them. A term such as r^3 means $r \times r \times r$ so it is 'three' letters together.

> the perimeter of a rectangle $P = 2l + 2w$
> the area of a square $A = x^2$
> the volume of a cuboid $V = lbh$

> the circumference of a circle $C = 2\pi r$
> the perimeter of a square $P = 4x$
> the perimeter of a triangle $P = a + b + c$

> the area of a circle $A = \pi r^2$
> the area of a rectangle $A = bh$
> the area of a triangle $A = \frac{1}{2}bh$

> the volume of a square $V = x^3$
> the volume of a cuboid $V = lbh$
> the volume of a cone $V = \frac{1}{3}\pi r^2 h$

- Scientists and mathematicians can check complicated **formulae** to see whether they are correct by testing **consistency**. They check that every term has the same **dimensions**.

- Each term in a formula must have the same dimensions.

- A formula with terms of different dimensions is said to be inconsistent and is not possible.

> Which of the following formulae are consistent? For each consistent formula, state whether it represents a length, an area or a volume.
>
> **a** $x + yz$ **b** $\pi r^3 + 2\pi r^2 h$ **c** $\dfrac{xy^2 + yx^2}{x}$ **d** $\dfrac{z(xy + r^2)}{q^2}$
>
> **a** inconsistent – it has length plus an area
> **b** consistent – represents a volume
> **c** consistent – represents an area
> **d** consistent – represents a length

Questions

Grade C

1 Some of the formulae below represent lengths, some represent areas and some represent volumes. Identify which is which.

  a πr **b** $x^2 y$ **c** pq
  d pqr **e** $a + b$ **f** $\pi r^2 h$

Grade B

2 State whether these formulae represent lengths (L), areas (A), volumes (V) or none of these (N).

  a $\frac{1}{2}bh + \pi r^2$ **b** $\dfrac{2p + 2q}{r}$ **c** $p^2 + q^3$
  d $\pi r^2(2h + r)$ **e** $a + ab + abc$ **f** $\pi(a + b + c)$

Vectors

A

Vectors

- A **vector** is a quantity that has both **magnitude** and **direction**.

- A vector can be represented by a straight line with an arrowhead pointing in the direction of the vector and the length representing the magnitude of the vector.

- Vectors are written as single letters in bold type, for example **a**, or as a pair of letters with an arrow above, for example $\vec{PQ}$.

The vector **p** shows a translation from A to B in the direction from A to B.
It can also be written as $\vec{AB}$.

A

Addition and subtraction of vectors

- Vectors can be added and subtracted, using the normal rules of algebra and numeracy. However, they also represent movements in space.

Top Tip!

If $\vec{OA}$ = **a** and
$\vec{OB}$ = **b**, then
$\vec{AB}$ = **b** – **a**.

a + **b** and **a** – **b** are shown in the diagrams opposite.

- When vectors relate to an origin, they are called **position vectors**.

In the grid below $\vec{OA}$ = **a**,
$\vec{OB}$ = **b**,
$\vec{OP}$ = 3**a** + 2**b**,
$\vec{OM}$ = **a** + 3**b**
and $\vec{GR}$ = 2**a** + **b**.

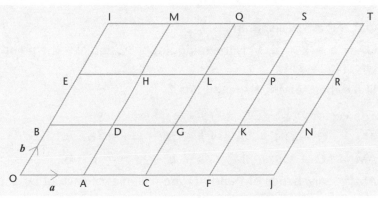

Questions

Grade A

1 Refer to the grid above.

  a Write down the following vectors in terms of **a** and **b**.

    i $\vec{EM}$ **ii** $\vec{DN}$ **iii** $\vec{DS}$
    iv $\vec{KH}$ **v** $\vec{RF}$

  b What is the relationship between the vectors $\vec{BH}$ and $\vec{AT}$?

  c What is the relationship between the vectors $\vec{BQ}$ and $\vec{RC}$?

SHAPE, SPACE AND MEASURES

Vector geometry

- **Vectors** can be used to prove results in geometry.

- In vector geometry problems, the key is to find a 'route' following vectors that you know or can find.

- When they form a 'route', vectors link together so that the last letter of one vector will be the first letter of the second vector.

$$\overrightarrow{AB} = \overrightarrow{AX} + \overrightarrow{XY} + \overrightarrow{YZ} + \overrightarrow{ZB}$$

- GCSE questions usually ask for a conclusion to be made. This consists of three possible answers.

    - The vectors are parallel. In this case the vectors are multiples of each other.

    - The vectors are on a straight line. In this case the vectors are multiples of each other and have a common point.

    - The vectors are opposite to each other. In this case the vectors will be negative multiples of each other.

- The opposite (negative) of any vector can be formed by reversing the letters.
$$\overrightarrow{AB} = -\overrightarrow{BA}$$

> **Top Tip!**
>
> GCSE questions always start with finding simple vectors which are needed later for more complex vectors. You should always be aware of using earlier answers.

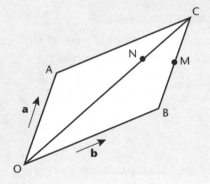

OACB is a parallelogram.

$\overrightarrow{OA} = \mathbf{a}$, $\overrightarrow{OB} = \mathbf{b}$, M is the midpoint of BC and N is the point such that ON : NC = 2 : 1.

Show that ANM is a straight line.

The vector $\overrightarrow{ON}$ is $\frac{2}{3}\overrightarrow{OC} = \frac{2}{3}(\mathbf{a} + \mathbf{b})$

$\overrightarrow{AN} = \overrightarrow{AO} + \overrightarrow{ON} = -\mathbf{a} + \frac{2}{3}(\mathbf{a} + \mathbf{b}) = -\frac{1}{3}\mathbf{a} + \frac{2}{3}\mathbf{b}$

$\overrightarrow{AM} = \overrightarrow{AO} + \overrightarrow{OB} + \overrightarrow{BM} = -\mathbf{a} + \mathbf{b} + \frac{1}{2}\mathbf{a} = -\frac{1}{2}\mathbf{a} + \mathbf{b}$

$\overrightarrow{AM} = \frac{3}{2}\overrightarrow{AN}$ hence $\overrightarrow{AM}$ and $\overrightarrow{AN}$ are multiples of each other and share a common point, A, so ANM is a straight line.

Questions

Grade A*

1 In the diagram above, the line OB is extended to a point P such that $\overrightarrow{OP} = 2\mathbf{b}$.

   a Write down the following vectors in terms of **a** and/or **b**.

   i $\overrightarrow{AB}$ **ii** $\overrightarrow{NC}$

   b Show that AMP is a straight line.

Shape, space and measures checklist

I can...

- [ ] find the area of a parallelogram, using the formula $A = bh$
- [ ] find the area of a trapezium, using the formula $\frac{1}{2}(a + b)h$
- [ ] find the area of a compound shape
- [ ] work out the formula for the perimeter, area or volume of simple shapes
- [ ] identify the planes of symmetry for 3-D shapes
- [ ] recognise and find alternate angles in parallel lines and a transversal
- [ ] recognise and find corresponding angles in parallel lines and a transversal
- [ ] recognise and find interior angles in parallel lines and a transversal
- [ ] use and recognise the properties of quadrilaterals
- [ ] find the exterior and interior angles of regular polygons
- [ ] understand the words 'sector' and 'segment' when used with circles
- [ ] calculate the circumference of a circle, giving the answer in terms of π if necessary
- [ ] calculate the area of a circle, giving the answer in terms of π if necessary
- [ ] recognise plan and elevation from isometric and other 3-D drawings
- [ ] translate a 2-D shape
- [ ] reflect a 2-D shape in lines of the form $y = a$ and $x = b$
- [ ] rotate a 2-D shape about the origin
- [ ] enlarge a 2-D shape by a whole number scale factor about the origin
- [ ] construct diagrams accurately using compasses, a protractor and a straight edge
- [ ] use the appropriate conversion factors to change imperial units to metric units and vice versa

You are working at (Grade D) level.

- [ ] work out the formula for the perimeter, area or volume of complex shapes
- [ ] work out whether an expression or formula represents a length, an area or a volume
- [ ] relate the exterior and interior angles in regular polygons to the number of sides
- [ ] find the area and perimeter of semicircles
- [ ] translate a 2-D shape, using a vector
- [ ] reflect a 2-D shape in the lines $y = x$ and $y = -x$
- [ ] rotate a 2-D shape about any point
- [ ] enlarge a 2-D shape by a fractional scale factor
- [ ] enlarge a 2-D shape about any centre
- [ ] construct perpendicular and angle bisectors
- [ ] construct an angle of 60°
- [ ] construct the perpendicular to a line from a point on the line and from a point to a line
- [ ] draw simple loci
- [ ] work out the surface area and volume of a prism
- [ ] work out the volume of a cylinder, using the formula $V = \pi r^2 h$
- [ ] find the density of a 3-D shape
- [ ] find the hypotenuse of a right-angled triangle, using Pythagoras' theorem

☐ find the short side of a right-angled triangle, using Pythagoras' theorem

☐ use Pythagoras' theorem to solve real-life problems

You are working at (Grade C) level.

☐ calculate the length of an arc and the area of a sector

☐ use trigonometric ratios to find angles and sides in right-angled triangles

☐ use trigonometry to solve real-life problems involving right-angled triangles

☐ use circle theorems to find angles, in circle problems

☐ use the conditions for congruency to identify congruent triangles

☐ enlarge a shape by a negative scale factor

☐ use similar triangles to find missing lengths

☐ identify whether a formula is dimensionally consistent, using dimensional analysis

☐ understand simple proofs such as the exterior angle of a triangle is equal to the sum of the opposite interior angles

You are working at (Grade B) level.

☐ calculate the surface area of cylinders, cones and spheres

☐ calculate the volume of cones and spheres

☐ solve 3-D problems using Pythagoras' theorem

☐ use the alternate segment theorem to find angles in circle problems

☐ prove that two triangles are congruent

☐ solve more complex loci problems

☐ solve problems using area and volume scale factors

☐ solve real-life problems, using similar triangles

☐ solve problems using addition and subtraction of vectors

☐ use the sine rule to solve non right-angled triangles

☐ use the cosine rule to solve non right-angled triangles

☐ use the rule $A = \frac{1}{2}ab\sin C$ to find the area of non right-angled triangles

You are working at (Grade A) level.

☐ calculate the volume and surface area of compound 3-D shapes

☐ use circle theorems to prove geometrical results

☐ solve more complex problems, using the proportionality of area and volume scale factors

☐ solve problems using vector geometry

☐ use the sine rule to solve more complex problems, involving right-angled and non right-angled triangles

☐ use the cosine rule to solve more complex problems, involving right-angled and non right-angled triangles

☐ solve 3-D problems using trigonometry

☐ find two angles between 0° and 360° for values of sine and cosine

☐ solve simple trigonometric equations where sine or cosine is the subject

☐ prove geometrical results with a logical and rigorous argument

You are working at (Grade A*) level.

Basic algebra

Top Tip!

Always use brackets and remember the rules for dealing with negative numbers.

Substitution

D

- Substitution means replacing letters in formulae and expressions with numbers.

- When replacing letters with numbers, use brackets to avoid problems with minus signs.

Work out the value of $ab + c$ if $a = -3$, $b = 4$ and $c = 5$.
$ab + c = (-3)(4) + (5) = -12 + 5 = -7$

Top Tip!

There is no need to show all the steps when expanding brackets as, usually, one mark is given for each correct term.

Expansion and simplification

D-C

- **Expand** in mathematics means **multiply out**.

  Expressions such as $4(z + 3)$ and $5x^2(x - 8)$ can be multiplied out.

- There is an **invisible multiplication sign** between the outside term and the opening bracket.

  $4(2x + 3)$ means $4 \times (2x + 3)$

- When expanding **brackets**, it is important to remember that the term outside the bracket is multiplied by each term inside the brackets.

  $4(2x + 3)$ means $4 \times (2x + 3) = 4 \times 2x + 4 \times 3 = 8x + 12$.
  You would normally just write $5x^2(x - 8) = 5x^3 - 40x^2$.

- When you are asked to **expand and simplify** an expression, it means expand any brackets and then simplify by collecting like terms.

  Expand and simplify $4(3 + m) - 5(2 - 3m)$.
  First, expand both brackets: $12 + 4m - 10 + 15m$
  Second, simplify: $2 + 19m$

Top Tip!

Do not try to expand and simplify in one go. If you try to do two things at once you will probably do one of them wrong.

Factorisation

D-C

- **Factorisation** is the opposite of expansion. Factorisation puts an expression back into the form $4(3x - 2)$.

- To factorise expressions, look for the **highest common factor** of both terms.

  $5x + 20$ has a common factor of 5 in each term,
  so $5x + 20 = 5 \times x + 5 \times 4 = 5(x + 4)$
  $4xy - 8x^2$ has a common factor of $4x$, so $4xy - 8x^2 = 4x \times y - 4x \times 2x = 4x(y - 2x)$

- Check your factorisation by multiplying out the final answer to check it goes back to what you started with.

Questions

Grade D

1 If $a = 3$, $b = -4$ and $c = 5$, find the value of each the following.
  a $ab + c$ **b** $a^2 + b^2$ **c** $2(a + 3b - c)$

Grade D

2 Expand the following expressions.
  a $3(x + 5)$ **b** $n(n - 7)$ **c** $3p^2(2p - 3q)$

Grade C

3 Expand and simplify the following expressions.
  a $2(x - 3) + 4(x + 3)$ **b** $8(x + 3y) + 2(x + 7y)$

Grade D

4 Factorise the following expressions.
  a $6x^2 - 9x$ **b** $2a^2b - 8ab + 6ab^2$

Linear equations

Solving linear equations

- An **equation** is formed when an expression is set equal to a number or another expression.
- A **linear equation** is one that only involves one **variable**.

$2x + 3 = 7$ and $5x + 8 = 3x - 2$ are both linear equations.

- **Solving** an equation means finding the value of the variable that makes it true.

Solve $2x + 3 = 7$. The value of x that makes this true is 2 because $2 \times 2 + 3 = 7$.

- There are four ways to solve equations. There is not much difference between them but **rearrangement** is the most efficient method.

Solve $6x + 5 = 14$.
Move the 5 across the equals sign to give: $6x = 14 - 5 = 9$
Divide both sides by 6 to give: $x = \frac{9}{6} \Rightarrow x = 1.5$

> **Top Tip!**
> This is called 'change sides, change signs', which means that plus becomes minus (and vice versa) and multiplication becomes division (and vice versa).

Solving equations with brackets

- When an equation contains **brackets** you should multiply out the brackets first and then solve the equation in the normal way.

Solve $4(x + 3) = 30$.
Multiply out the brackets to give: $4x + 12 = 30$
Move the 12 across the equals sign to give: $4x = 30 - 12 = 18$
Divide both sides by 4 to give: $x = 4.5$

> **Top Tip!**
> Always check your answer works in the original equation.

Equations with the variable on both sides of the equals sign

- When a letter appears on **both sides** of an equation, use the 'change sides, change signs' rule.

- Use the rule to collect all the terms containing the variable on the left-hand side of the equals signs and all number terms on the right-hand side.

Rearranging $5x - 3 = 2x + 12$ gives $5x - 2x = 12 + 3$.

> **Top Tip!**
> Be careful when moving terms across the equals sign and remember to change the signs from plus to minus and vice versa.

- After the equation is rearranged, the terms are collected together and the equation is solved in the usual way.

$5x - 3 = 2x + 12$ is rearranged to give $5x - 2x = 12 + 3 \Rightarrow 3x = 15 \Rightarrow x = 5$

Questions

Grades D-C

1 Solve these equations.

a $3x - 5 = 4$ **b** $2m + 8 = 6$ **c** $\frac{n}{3} = 4$
d $8x + 5 = 9$ **e** $\frac{x-3}{5} = 2$ **f** $5x - 3 = 7$
g $\frac{y+2}{5} = 3$ **h** $\frac{x}{7} - 3 = 1$ **i** $8 - 2x = 7$

Grade D

2 Solve these equations.
 a $3(x - 5) = 12$ **b** $2(m - 3) = 6$ **c** $3(x + 5) = 9$

d $2(x - 3) = 2$ **e** $6(x - 1) = 9$ **f** $4(y + 2) = 2$

Grade D

3 Solve these equations.

 a $3x - 2 = x + 12$ **b** $5y + 5 = 2y + 11$
 c $8x - 1 = 5x + 8$ **d** $7x - 3 = 2x - 8$
 e $6x - 2 = x - 1$ **f** $3y + 7 = y + 2$

Remember: You must revise all content from Grade E to the level that you are currently working at.

C

Equations with brackets and the variable on both sides

- When an equation contains brackets and variables on **both sides**, always expand the brackets first.

  Expanding $4(x - 2) = 2(x + 3)$ gives $4x - 8 = 2x + 6$.

- After the brackets are expanded, the equation is rearranged, the terms are collected and the equation is solved in the usual way.

  Expanding $4(x - 2) = 2(x + 3)$ gives $4x - 8 = 2x + 6 \Rightarrow 4x - 2x = 8 + 6 \Rightarrow 2x = 14 \Rightarrow x = 7$.

D-C

Setting up equations

- Many **real-life** problems can be solved by using equations to model them.

Top Tip!

Use the letter x when setting up an equation unless you are given another letter to use.

The angles in a triangle are given by $2x$, $3x$ and $4x$. What is the largest angle in the triangle?

You know the angles in a triangle add up to 180°, so:

$2x + 3x + 4x = 180$

This simplifies to: $9x = 180°$

Dividing both sides by 9 gives: $x = 20$

So the largest angle is $4x = 4 \times 20 = 80°$.

Fred is 29 years older than his daughter, Freda. Together their ages add up to 47.

Using the letter x to represent Freda's age, write down an expression for Fred's age.

Set up an equation in x and solve it to find Freda's age.

Fred is $x + 29$, $x + x + 29 = 47 \Rightarrow 2x + 29 = 47 \Rightarrow 2x = 18 \Rightarrow x = 9$

So Freda is 9 years old.

Questions

Grade C

1 Solve each of the following.

   a $5(x - 3) = 2(x + 6)$ **b** $6(x + 1) = 2(x - 3)$

   c $3(x + 5) = 2(x + 10)$ **d** $7(x - 2) = 3(x + 4)$

Grade D

2 The diagram shows a rectangle.

   a What is the value of x?

   b What is the value of y?

Grade D

3 A family has x bottles of milk delivered every day from Monday to Friday and 7 bottles delivered on Saturday. There is no delivery on Sunday. In total they have 22 bottles delivered each week.

   a Use the above information to set up an equation in x.

   b Solve your equation to find the value of x.

Grade C

4 Asil thought of a number. He divided it by 2 then added 7. The result was 6 more than the number he first thought of.

   a Use the above information to set up an equation, using x to represent the number Asil thought of.

   b Solve your equation to find the value of x.

Trial and improvement

Trial and improvement

- Some equations **cannot be solved** simply by using algebraic methods.

- A **numerical method** for solving these equations is called **trial and improvement**.

- After an initial guess (**the trial**), an answer is calculated and compared to the required answer.

- A better guess (**the improvement**) is then made.

- This process is **repeated** until the answer is within a given accuracy.

- Normally the first thing to do is find **two whole numbers** between which the answer lies.

- Then find **two decimal numbers**, **each with one decimal place**, between which the answer lies.

- Finally, test the **mid-way value** to see which of the two numbers is closer.

Use trial and improvement to solve the equation $x^3 + 2x = 52$, giving your answer to 1 decimal place.

Try $x = 4$ $4^3 + 2 \times 4 = 72$ Too high – next trial needs to be smaller

Try $x = 3$ $3^3 + 2 \times 3 = 33$ Too low

So we now know that the solution lies between $x = 3$ and $x = 4$.

Try $x = 3.5$ $3.5^3 + 2 \times 3.5 = 49.875$ Close, but too low

Try $x = 3.6$ $3.6^3 + 2 \times 3.6 = 53.856$ Close, but too high

So we know the solution lies between $x = 3.5$ and $x = 3.6$.

To see which is closest, try $x = 3.55$ $3.55^3 + 2 \times 3.55 = 51.838\,875$, which is just too low.

Hence, the answer, correct to one decimal place, is 3.6.

- You could set out the working in a table.

Guess	$x^3 + 2x$	Comment
4	72	Too high
3	33	Too low
3.5	49.875	Too low
3.6	53.856	Too high
3.55	51.838 875	Too low

Questions

Grade C

1 Use trial and improvement to find the solution to $x^3 - 2x = 100$.
Give your answer to 1 decimal place.

Grade C

2 Use trial and improvement to find the solution to $x^3 + x = 20$.
Give your answer to 1 decimal place.

Simultaneous equations

Simultaneous equations

- **Simultaneous equations** are two or more equations for which we want to find the same solution. These are normally linear equations involving **two variables**.

  $2x + y = 10$ has many solutions: $x = 4, y = 2$; $x = 3, y = 4$; ...
  and $3x - y = 5$ has many solutions: $x = 4, y = 7$; $x = 3, y = 4$; ...
  Only one solution, $x = 3$ and $y = 4$, satisfies both equations at the same time.

- Simultaneous equations can be solved by one of two methods:
  - the **elimination** method
  - the **substitution** method

> **Top Tip!**
> Always label the equations so your method can be explained.

- Follow these steps to use the elimination method to solve simultaneous equations.
  - Step 1: **Balance** the **coefficients** of one of the **variables**
  - Step 2: Add or subtract the equations to **eliminate** this variable
  - Step 3: Solve the resulting linear equation to find the value of the other variable
  - Step 4: **Substitute** the value back into one of the previous equations
  - Step 5: Solve this equation find the value of the second variable

  Solve the equations $4x + 3y = 27$ and $5x - 2y = 5$.
     $4x + 3y = 27$ (1)
     $5x - 2y = 5$ (2)
  Balance the coefficent of x by multiplying (1) by 5 and (2) by 4:
     (1) $\times$ 5 $20x + 15y = 135$ (3)
     (2) $\times$ 4 $20x - 8y = 20$ (4)
  Subtract equation (4) from equation (3):
     (3) $-$ (4) $23y = 115 \Rightarrow y = 5$
  Substitute $y = 5$ into (1):
     $4x + 15 = 27 \Rightarrow 4x = 12 \Rightarrow x = 3$

> **Top Tip!**
> If the balanced terms have the same sign subtract the equations. If they have different signs add the equations.

- The **substitution method** works by substituting one equation into the other.

  Solve the equations $6x + y = 9$ and $y = 3x - 9$.
     $6x + y = 9$ (1)
     $y = 3x - 9$ (2)
  Substitute the right-hand side of equation (2) into equation (1):
     $6x + (3x - 9) = 9$
  Expand and solve the equation:
     $9x - 9 = 9 \Rightarrow 9x = 18 \Rightarrow x = 2$
  Substitute $x = 2$ into equation (1):
     $6 \times 2 + y = 9 \Rightarrow y = -3$

> **Top Tip!**
> Always test the values in the original equations.

Questions

Grade B

1 Solve these simultaneous equations.

a $3x + y = 7$ **b** $3x + 5y = 15$ **c** $2x + 5y = 15$
   $2x + y = 4$ $x + 3y = 7$ $3x - 2y = 13$

Grade B

2 Solve the equations $4x - y = 17$ and $x = 5 + y$.

Simultaneous equations and formulae

Setting up simultaneous equations

- Many real-life situations can be solved by simultaneous equations.

Four cakes and three cups of tea cost £6.40.

Three cakes and two cups of tea cost £4.50.

How much will it cost to buy three cakes and four cups of tea?

Let x be the cost of a cake.

Let y be the cost of a cup of tea.

$4x + 3y = 640$ (1)

$3x + 2y = 450$ (2)

Solve these equations using the method already demonstrated:

(1) × 3 $12x + 9y = 1920$ (3)

(2) × 4 $12x + 8y = 1800$ (4)

(3) – (4) $y = 120$

Substitute into (1): $4x + 360 = 640 \Rightarrow 4x = 280 \Rightarrow x = 70$

The cost of three cakes and four cups of tea will be $3 \times 70 + 4 \times 120 = £6.90$.

A

Rearranging formulae

C-B

- The subject of a formula is the **variable** (letter) in the formula that stands on its own, usually on the left-hand side of the equals sign.

x is the subject of each of these formulae.

$x = 5t + 4$ $x = 4(2y - 7)$ $x = \dfrac{1}{t}$

Top Tip!

Remember the rules 'change sides, change signs' and 'what you do to one side, you do to the other'.

- To change the subject of a formula, you have to **rearrange** the formula to get the required variable on the left-hand side.

- To rearrange a formula, use the same rules as for solving equations.

- The main difference is that, when rearranging a formula, each step gives an algebraic expression rather than a numerical value.

Make m the subject of $y = 3m - t$.

Add t to both sides: $y + t = 3m$

Reverse the formula: $3m = y + t$

Divide both sides by 3: $m = \dfrac{y + t}{3}$

Questions

Grade A

1 Two families visit the zoo.
The Collins family pay £30 for two adults and three children.
The Gordon family pay £27.50 for one adult and four children.

a How much does it cost for one adult?

b How much does it cost for one child?

Grade C

2 Rearrange the following formulae to make x the subject.

a $T = 4x$ **b** $y = 2x + 3$ **c** $P = 2t + x$

d $y = \dfrac{x}{5}$ **e** $A = mx + y$ **f** $S = 2\pi x^2$

Algebra 2

Quadratic expansion

- A **quadratic expression** is one in which the highest power of any term is 2.

  x^2, $4y^2 + 3y$ and $6x^2 - 2x + 1$ are quadratic expressions.

- An expression such as $(x + 2)(x - 3)$ can be **expanded** to give a quadratic expression.

- When an expression such as $(x + 2)(x - 3)$ is multiplied out, it is called **quadratic expansion**.

- There are three methods for quadratic expansion.

  - **Splitting the brackets** The terms inside the first brackets are split and used to multiply the terms in the second brackets.

    Expand $(x + 2)(x + 5)$.
    $(x + 2)(x + 5) = x(x + 5) + 2(x + 5) = x^2 + 5x + 2x + 10 = x^2 + 7x + 10$

  - **FOIL** FOIL stands for First, Outer, Inner and Last. This is the order in which terms are multiplied.

    Expand $(x - 3)(x + 4)$.
    First terms give: $x \times x = x^2$. Outer terms give: $x \times 4 = 4x$.
    Inner terms give: $-3 \times x = -3x$. Last terms give: $-3 \times + 4 = -12$.
    $(x - 3)(x + 4) = x^2 + 4x - 3x - 12 = x^2 + x - 12$

  - **Box method** This is similar to the box method used for long multiplication.

    Expand $(x - 4)(x - 2)$.

$\times$	x	$- 4$
x	x^2	$- 4x$
$- 2$	$- 2x$	$+ 8$

    $= x^2 - 4x - 2x + 8$
    $= x^2 - 6x + 8$

> **Top Tip!**
> When you multiply out a quadratic expansion, there will always be four terms and two of these terms will combine together.

> **Top Tip!**
> Be careful with the signs. This is the main place where marks are lost in examination questions involving the expansion of brackets. Remember: $-2 \times -4 = +8$.

Squaring brackets

- When you are asked to square a term in brackets, such as $(x + 3)^2$, you must write down the bracket twice, $(x + 3)(x + 3)$, and then use whichever method you prefer to expand the brackets.

  Expand $(x + 3)^2$.
  $(x + 3)^2 = (x + 3)(x + 3) = x^2 + 3x + 3x + 9 = x^2 + 6x + 9$

Questions

Grade C

1 Expand the following expressions.

  a $(x - 1)(x + 3)$ **b** $(m + 2)(m - 6)$

  c $(n + 1)(n - 2)$ **d** $(x + 5)(x + 1)$

  e $(x - 3)(x - 3)$ **f** $(x + 3)(x + 7)$

Grade C

2 Expand the following squares.

  a $(x + 3)^2$ **b** $(x - 2)^2$

Grade C

3 Expand the following expressions.

  a $(x - 1)(x + 1)$ **b** $(m + 2)(m - 2)$

Factorising quadratic expressions

Factorising a quadratic with a unit coefficient of x^2

- **Factorisation** involves putting a quadratic expression back into its **brackets** (if possible).

- When an expression such as $x^2 + ax + b$ is factorised, it is always of the form $(x + p)(x + q)$.

  $x^2 + ax + b = (x + p)(x + q)$, where $p + q = a$ and $pq = b$.

- You can pick up clues about the signs in the brackets from the signs in the quadratic equation.

  $x^2 + ax + b = (x + p)(x + q)$ $x^2 - ax + b = (x - p)(x - q)$ $x^2 \pm ax - b = (x + p)(x - q)$

  Factorise $x^2 - x - 2$.

  You know that the factorisation must be of the form $(x + p)(x + q)$, so look for two numbers, p and q, such that $p + q = -1$ and $pq = -2$. So, $x^2 - x - 2 = (x + 1)(x - 2)$.

- **Factorising** a quadratic expression with a coefficient in front of the x^2 term is more complicated.

- When an expression such as $ax^2 + bx + c$ is factorised, it is always of the form $(mx + p)(nx + q)$.

  $ax^2 + bx + c = (mx + p)(nx + q)$, where $mn = a$ and $pq = c$.

  Factorise $3x^2 + 5x - 2$.

  You know the factorisation must be of the form $(mx + p)(nx + q)$, so you need to find a combination of m, n, p and q that will give the right number of xs. Considering the factors of 2 and 3, the solution is $3x^2 + 5x - 2 = (3x - 1)(x + 2)$.

B-A

Difference of two squares

- Multiplying out $(x + a)(x - a)$ gives $x^2 - a^2$ because the x terms cancel each other out.

- This expansion is called the **difference of two squares**. It can be used to factorise certain special quadratic expressions.

- Reversing the above, $x^2 - a^2 = (x + a)(x - a)$, where a^2 is a square number.

  Factorise $x^2 - 9$.

  Recognise the difference of two squares, x^2 and 3^2. So it factorises to $(x + 3)(x - 3)$.

B

Solving quadratic equations by factorisation

- Once a quadratic expression has been factorised, it can be used to solve the equivalent quadratic equation.

  Solve $x^2 + 6x + 8 = 0$.

  This factorises into $(x + 2)(x + 4) = 0$. Hence, either $x + 2 = 0 \Rightarrow x = -2$ or $x + 4 = 0 \Rightarrow x = -4$. So the solution is $x = -2$ or $x = -4$.

Top Tip!

The equation must be in the form as $ax^2 + bx + c = 0$ before it can be solved.

B

Questions

(Grade B)

1 Factorise these expressions.

  a $x^2 + 2x - 3$ **b** $x^2 - 5x + 6$

(Grade B)

2 Factorise these expressions.

  a $x^2 - 16$ **b** $x^2 - 36$

(Grade A)

3 Factorise these expressions.

  a $2x^2 - 9x - 5$ **b** $6x^2 + 5x - 6$

(Grade B)

4 First factorise, then solve these equations.

  a $x^2 - 2x - 3 = 0$ **b** $x^2 + 3x - 10 = 0$

Solving quadratic equations

Solving quadratics of the form $ax^2 + bx + c = 0$

- The general quadratic equation is of the form $ax^2 + bx + c = 0$, where a, b and c are positive or negative whole numbers.
- You can solve a general quadratic equation by factorisation, the quadratic formula or completing the square.

Solving a quadratic equation by factorisation

- **Factorise** the quadratic and solve the simple **linear equation** that results when you put each bracket **equal to zero**.

  Solve $6x^2 + 5x - 4 = 0$. This factorises into $(2x - 1)(3x + 4) = 0$.
  Hence, either $2x - 1 = 0$ or $3x + 4 = 0 \Rightarrow x = \frac{1}{2}$ or $\Rightarrow x = \frac{-4}{3}$
  So the solution is $x = \frac{1}{2}$ or $x = \frac{-4}{3}$.

- When b is zero, the equation can be rearranged to the form $ax^2 = -c$.

  Solve $4x^2 - 9 = 0$. $4x^2 - 9 = 0 \Rightarrow 4x^2 = 9 \Rightarrow x^2 = \frac{9}{4} \Rightarrow x = \sqrt{\frac{9}{4}} = \frac{3}{2}$

- When c is zero, the equation is solved by taking out a common factor.

  Solve $5x^2 - 15x = 0$. $5x^2 - 15x = 0 \Rightarrow 5x(x - 3) = 0 \Rightarrow x = 0$ or $x = 3$

Top Tip!
When c is zero, one of the solutions is always $x = 0$.

Solving the general quadratic by the quadratic formula

- Many quadratic equations cannot be solved by factorisation because they do not have simple factors.
- One way to solve this type of equation is to use the **quadratic formula**. This formula can solve any soluble quadratic equation.
- The solutions of the equation $ax^2 + bx + c = 0$ are given by:
  $$x = \frac{-b \pm \sqrt{b^2 - 4ac}}{2a}$$

  Solve $4x^2 + 3x - 2 = 0$. Take the quadratic formula and put $a = 4$, $b = 3$ and $c = -2$:
  $$x = \frac{-3 \pm \sqrt{3^2 - 4(4)(-2)}}{2(4)}$$

  Note that the values for a, b and c have been put into the formula in brackets to avoid mistakes in the calculation.
  $$x = \frac{-3 \pm \sqrt{9 + 32}}{8} = \frac{-3 \pm \sqrt{41}}{8} \Rightarrow x = 0.43 \text{ or } x = -1.18$$

Top Tip!
The quadratic formula is given on the formula sheet that is included with the examination. However, it is better to learn it because many students miscopy the formula.

Questions

Grade A
1 Solve these equations by factorisation.
  a $2x^2 + x - 6 = 0$ **b** $12x^2 + 17x + 6 = 0$

Grade A
2 Solve these equations. Give your answer in surd form if appropriate.
  a $2x^2 - 7 = 0$ **b** $4x^2 - 12x = 0$

Grade A
3 Solve $2x^2 - 3x - 1 = 0$, using the quadratic formula. Give your answer to 2 decimal places.

Remember: You must revise all content from Grade E to the level that you are currently working at.

Solving a quadratic equation by completing the square

A–A*

- The third method for solving quadratic equations is **completing the square**.

- This method can be used to give answers to a specified number of decimal places or to leave answers in **surd** form.

- The expansion of $(x + a)^2$ is $x^2 + 2ax + a^2$ and rearranging $x^2 + 2ax + a^2 = (x + a)^2$ gives $x^2 + 2ax = (x + a)^2 - a^2$.

- This is the basic principle behind completing the square.

> Rewrite $x^2 + 6x - 5$ in the form $(x + a)^2 - b$.
> We note that: $x^2 + 6x = (x + 3)^2 - 9$
> So, we have: $x^2 + 6x - 5 = (x + 3)^2 - 9 - 5 = (x + 3)^2 - 14$

- Once a quadratic equation has been rewritten by completing the square, it can be used to solve the equation.

> Solve $x^2 + 8x - 5 = 0$.
> First rewrite the left-hand side by the method of completing the square.
> We note that: $x^2 + 8x = (x + 4)^2 - 16$
> So we have: $x^2 + 8x - 5 = (x + 4)^2 - 16 - 5 = (x + 4)^2 - 21$
> So $x^2 + 8x - 5 = 0 \Rightarrow (x + 4)^2 - 21 = 0$
> Rearranging gives: $(x + 4)^2 = 21$
> Taking the square root of both sides gives:
> $x + 4 = \pm \sqrt{21}$
> $\Rightarrow x = -4 \pm \sqrt{21}$
> This answer could be left in surd form or evaluated to a given accuracy.

> **Top Tip!**
> You will notice that the number inside the bracket is half the coefficient of x. In GCSE examinations, the coefficient is always even to avoid problems with fractions.

Quadratic equations with no solution

A*

- The quantity $b^2 - 4ac$ in the quadratic formula is known as the **discriminant**.
- When $b^2 - 4ac$ is positive, the equation has two roots or solution.
- When $b^2 - 4ac$ is zero, the equation has one repeated root or one solutions.
- When $b^2 - 4ac$ is negative, the equation has no roots and is not soluble.

> Find the discriminant $b^2 - 4ac$ of the equation $x^2 - 3x + 5 = 0$ and explain what the result tells you.
> $b^2 - 4ac = (-3)^2 - 4(1)(5) = 9 - 20 = -11$, so the equation has no roots.

> **Top Tip!**
> You will not be asked about this in GCSE examinations so, if this happens, you have made a mistake and should check your working.

Questions

Grade A

1 Write an equivalent expression in the form $(x \pm a)^2 - b$.

  a $x^2 + 6x - 1$ **b** $x^2 - 8x - 3$

Grade A*

2 Solve these equations by completing the square. Leave your answers in surd form.

  a $x^2 + 4x - 1 = 0$ **b** $x^2 - 6x + 2 = 0$

Grade A*

3 a As part of her coursework, Mary worked out the quadratic equation $x^2 + 6x + 10 = 0$. She could not solve the equation. Explain why.

  b The equation Mary should have found was $x^2 + 6x - 10 = 0$. Use the method of completing the square to solve this equation.

A

Using the quadratic formula without a calculator

- GCSE examination questions do not specify what method you should use to solve a quadratic equation.

- In the non-calculator paper, you could be asked to solve a quadratic equation that does not factorise. In this case, you would be asked to leave your answer in root or surd form. You could use completing the square, but sometimes the quadratic formula is easier to use.

Solve $x^2 - x - 11 = 0$.

This equation does not factorise and the coefficient of x is odd so completing the square will give fractional values.

Putting $a = 1$, $b = -1$ and $c = -11$ into the quadratic formula gives:

$$x = \frac{-(-1) \pm \sqrt{(-1)^2 - 4(1)(-11)}}{2(1)} = \frac{1 \pm \sqrt{45}}{2} = \frac{1 \pm 3\sqrt{5}}{2}$$

A*

Solving problems with quadratic equations

- Many real-life problems can be solved by modelling with quadratic equations.

The right-angled triangle shown has sides of $x + 1$, $4x$ and $5x - 1$. Find the value of x.

As the triangle is right-angled, the sides must obey Pythagoras' theorem.

So, $(5x - 1)^2 = (4x)^2 + (x + 1)^2$

$\Rightarrow 25x^2 - 10x + 1 = 16x^2 + x^2 + 2x + 1$

$\Rightarrow 8x^2 - 12x = 0$

$\Rightarrow 4x(2x - 3) = 0$

$\Rightarrow x = 0$ or $x = 1.5$

x cannot be 0, so $x = 1.5$.

Top Tip!

The quadratic equation you end up with will always factorise. If it does not, go back and check your working.

Questions

Grade A

1 Use the quadratic formula to solve these equations. Leave your answers in surd form.

   a $x^2 - 3x - 1 = 0$ **b** $2x^2 + 6x - 1 = 0$

Grade A*

2 a Show that $x - \frac{6}{x} = 1$ can be rearranged to $x^2 - x - 6 = 0$.

   b Solve $x^2 - x - 6 = 0$.

Grade A*

3 The sides of a rectangle are $(3x - 2)$ m and $(2x + 3)$ m. The area of the rectangle is 28 m^2. What is the value of x?

Real-life graphs

Travel graphs

- A **travel graph** shows information about a journey.

- Travel graphs are also known as **distance-time** graphs.

- Travel graphs show the main features of a journey and use **average speeds**, which is why the lines in them are straight.

This graph shows a car journey from Barnsley to Nottingham and back.

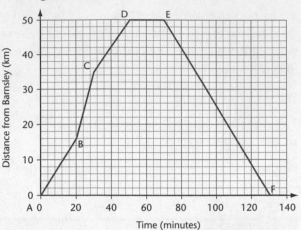

- In **reality**, vehicles do not travel at constant speeds.

- The average speed is given by:

$$\text{average speed} = \frac{\text{total distance travelled}}{\text{total time taken}}$$

- In a travel graph, the **steeper** the line, the **faster** the vehicle is travelling.

Top Tip!

When asked for an average speed, give the answer in kilometres per hour (km/h) or miles per hour (mph).

Real-life graphs

- Some situations can lead to unusual graphs.

This graph shows the depth of water in a conical flask as it is filled from a tap delivering water at a steady rate.

Questions

Grade D

1 Refer to the travel graph above.
  a After how many minutes was the car 16 kilometres from Barnsley?
  b What happened between points D and E?
  c Between which two points was the car travelling fastest?
  d What was the average speed for the part of the journey between C and D?

Grade B

2 Draw a graph of the depth of water in these containers as they are filled steadily.
  a   **b**

Trigonometry

2-D trigonometric problems

- Trigonometry can be used to solve more complicated problems than the straightforward right-angled triangles revised on pages 37–30.

Work out the area of this triangle.

Draw the perpendicular from A to BC.

Label the point where it meets CB, X.

Using triangle CXA, AX = 8 × sin 50° = 6.13 cm

Area = $\frac{1}{2}bh$ = $\frac{1}{2}$ × CB × AX = $\frac{1}{2}$ × 12 × 6.13 = 36.8 cm^2

A building has a ledge three-quarters of the way up, along the front, as shown in the diagram (seen from the side).

Top Tip!

Always redraw any relevant triangles and put on the known information.

The angle of elevation from a point X, 100 metres away from the building, to the ledge is 30°.

The angle of elevation of the top of the building from the same point is 35°.

Work out: **a** the height of the ledge, h

b the height of the building, t

c the width of the ledge, w

a Using the triangle opposite:

$h = 100 × \tan 30° = 57.7$ m

b h is three-quarters of t, so $t = 77.0$ m

c Using the triangle opposite:

$x = 77 ÷ \tan 35° = 110$ m

so the width of the ledge is 10 m.

Questions

Grade A

1 Refer to the second triangle in the first example box above. Work out:

a CX   **b** XB   **c** angle ABX

Grade A*

2 Refer to the diagram of the building above. The ledge is half the width of the building. An aerial that is 8 metres high is erected at the rear of the building. What is the angle of elevation of the top of the aerial from the point X?

3-D trigonometry

3-D trigonometric problems

- On pages 35–36, you revised solving 3-D problems by Pythagoras' theorem.
- You need to follow similar steps when solving 3-D problems by trigonometry.
  - Step 1: Identify the right-angled triangle that includes the required information
  - Step 2: Redraw this triangle as a separate right-angled triangle and label with the given information and the information to be found
  - Step 3: Work out the required value by trigonometry or identify another right-angled triangle if necessary
  - Step 4: Solve the problem and round to a suitable degree of accuracy

This cuboid has dimensions 10 cm by 15 cm by 25 cm.

M is the midpoint of FG. Calculate angle AMB.

First, identify a right-angled triangle that contains angle AMB and draw it.

This gives △AMB which contains AB = 10 cm
and two lengths that you do not know, MB and AM.

You need to find either MB or AM.

MB will provide an easier calculation, so let MB = x.

Next identify a right-angled triangle that contains MB and draw it.

This gives isosceles triangle BMC.

Using Pythagoras' theorem:

$$MB = \sqrt{25^2 + 7.5^2} = \sqrt{681.25} = 26.1 \text{ cm}$$

We can now go back to △AMB and use tangent to work out the angle.

$$\angle AMB = \tan^{-1}(10 \div 26.1) = \tan^{-1} 0.383 = 21°$$

Questions

Grade A*

1 Refer to the cuboid above.

  a Calculate angle AGB.

  b Calculate angle GEC.

Trigonometric ratios of angles from 0 to 360°

Trigonometric ratio of angles between 0° and 360°

- All the angles that you have revised with sine and cosine up until now have been less than 90°.

- Many triangles have obtuse angles so you will need to be able to find the sine and cosine values of angles greater than 90°.

- All angles, no matter what size, have sine, cosine and tangent values.

- This section revises sine and cosine values of angles from 0° to 360°.

Sine and cosine values of angles from 0° to 360°

- The graphs below shows sin x from 0° to 360° and cos x from 0° to 360°.

Sine x Cosine x

- Note the following facts.

  - Each and every value of sine and cosine between –1 and 1 gives two possible angles between 0° and 360°

  - When the value of sine is positive, both angles are between 0° and 180°

  - When the value of sine is negative, both angles are between 180° and 360°

  - When the value of cosine is positive, one angle is between 0° and 90° and the other is between 270° and 360°

  - When the value of cosine is negative, both angles are between 90° and 270°

Top Tip!

If you use your calculator to find sin⁻¹ –0.2, you will get a negative answer. Check sin 191.5 and sin 348.5 you get –0.2.

Find two angles with a sine of –0.2.
You know that both angles are between 180° and 360°.
Use your calculator to find $\sin^{-1}$ 0.2: –11.5°
So the angles are 180° + 11.5° and 360° – 11.5° which gives 191.5° and 348.5°.

Find the angles with a cosine of –0.8.
You know that both angles are between 90° and 270°.
Use your calculator to find $\cos^{-1}$ 0.8: 36.9°
From the graph, and using symmetry, the angles are:
180° – 36.9° = 143.1°, 180° + 36.9° = 216.9°
So the two angles are 143.1° and 216.9°.

Top Tip!

Use your calculator to check cos 143.1 and cos 216.9 = –0.8.

Questions

Grade A*

1 State the two angles between 0° and 360° with each of these values.

  a a sine of 0.4   **b** a cosine of 0.1   **c** a sine of –0.7   **d** a cosine of –0.4

Sine rule

Solving any triangle

- Up until now you have only revised right-angled triangles.

- Any triangle has six elements: three sides and three angles.

- To solve a triangle, that is to find any unknown angle or side, you need to know at least three of the elements. Any combination, except for three angles, is enough to work out the rest.

- You can use either the sine rule or the cosine rule to solve a triangle that does not contain a right angle.

A

The sine rule

- This is the proof of the sine rule.

  Take a triangle, ABC, and draw the perpendicular from A to the opposite side BC. Label the point of intersection D and the line h.

  From right-angled triangle ADB, $h = c \sin B$

  From right-angled triangle ADC, $h = b \sin C$

  Therefore, $c \sin B = b \sin C$ which can be rearranged to give $\frac{c}{\sin C} = \frac{b}{\sin B}$

  By algebraic symmetry, we see that $\frac{a}{\sin A} = \frac{c}{\sin C}$ and $\frac{a}{\sin A} = \frac{b}{\sin B}$

  This can combined and inverted to give $\frac{\sin A}{a} = \frac{\sin B}{b} = \frac{\sin C}{c}$

  Note that the convention is to label angles with capital letters and sides with lower case letters, with the corresponding letters opposite each other.

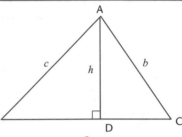

Top Tip!

Only the version of the sine rule with the *sides on top* is given on the GCSE formula sheet. If you need to find an angle, remember to invert it.

A

In triangle ABC, find the value of x.

Use the sine rule with sides on top, which gives $\frac{x}{\sin 48°} = \frac{12}{\sin 63°}$

$\Rightarrow x = \frac{12 \sin 48°}{\sin 63°} = 10.0$ cm (3 significant figures)

The ambiguous case

- When you find an angle, you end up with a value of sine, such as $\sin x = 0.9$.

- As you know from page 69, this gives two angles between 0° and 180° giving two possible triangles.

A

Top Tip!

Examiners will not try to catch you out with this. They will indicate clearly, either in writing or with the aid of a diagram, what is required.

or

Questions

(Grade A)

1 Refer to the triangle in the tinted box above.

   a Find the value of angle B.

   b Use the sine rule to find the length of side AC.

(Grade A)

2 Refer to the triangle with sides of 9 cm and 7 cm and an angle of 40° above. Use the sine rule to find the two possible values of angle C.

Cosine rule

The cosine rule

- This is the proof of the cosine rule.

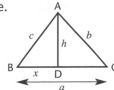

Take a triangle ABC and draw the perpendicular from A to the opposite side BC. Label the point of intersection D and the line AD, h.

Using Pythagoras on triangle BDA, $h^2 = c^2 - x^2$

Using Pythagoras on triangle ADC, $h^2 = b^2 - (a - x)^2$

Therefore, $c^2 - x^2 = b^2 - (a - x)^2$

$$c^2 - x^2 = b^2 - a^2 + 2ax - x^2$$

$$\Rightarrow c^2 = b^2 - a^2 + 2ax$$

From triangle BDA, $x = c \cos B$

Hence, $c^2 = b^2 - a^2 + 2ac \cos B$

Rearranging gives $b^2 = a^2 + c^2 - 2ac \cos B$

By algebraic symmetry $a^2 = b^2 + c^2 - 2bc \cos A$ and $c^2 = a^2 + b^2 - 2ab \cos C$

$a^2 = b^2 + c^2 - 2bc \cos A$ can be rearranged to give

$$\cos A = \frac{b^2 + c^2 - a^2}{2bc}$$

Find x in this triangle.

By the cosine rule, $x^2 = 8^2 + 10^2 - 2 \times 8 \times 10 \times \cos 85°$

$$x^2 = 150.1$$

$$\Rightarrow x = 12.2 \text{ cm}$$

Find x in this triangle.

By the cosine rule, $\cos x = \dfrac{11^2 + 13^2 - 20^2}{2 \times 11 \times 13}$

$$\cos x = -0.3846...$$

$$\Rightarrow x = 112.6°$$

Questions

Grade A

1 Find the length x in this triangle. Give your answer to 1 decimal place.

Grade A

2 Find angle x in this triangle.

Solving triangles

Which rule to use

- When solving triangle problems, there are only four possible situations that can occur, each of which can be solved completely in three stages.

- **Two sides and the included angle**

  1 Use cosine rule to find the third side

  2 Use the sine rule to find either of the other two angles

  3 Use the sum of the angles in a triangle to find the third angle

- **Two angles and a side**

  1 Use the angle sum of the angles in a triangle to find the third angle

  2 Use the sine rule to find either of the other two sides

  3 Use the sine rule to find the third side

- **Three sides**

  1 Use the cosine rule to find any one of the angles

  2 Use the sine rule to find either of the other two angles

  3 Use the sum of the angles in a triangle to find the third angle

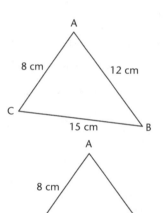

- **Two sides and a non-included angle**

  1 Use the sine rule to find the two possible values of the appropriate angle. Use information in the question to decide if the angle is acute or obtuse and select the appropriate value.

  2 Use the sum of the angles in a triangle to find the third angle.

  3 Use the sine rule to find the third side.

A

Using sine to find the area of a triangle

- If you know two sides and the included angle of a triangle, you can use the formula: **area** $= \frac{1}{2} \times a \times b \times \sin C$ to calculate its area.

Find the area of triangle ABC.

Area $= \frac{1}{2}ab \sin C$

Area $= \frac{1}{2} \times 8 \times 10 \times \sin 42 = 26.8$ cm^2 (3 significant figures)

A

Questions

1 a Find the side BC in the first triangle above.

  b Find the side BC in the second triangle above.

2 a Find angle A in the third triangle above.

  b Find angle A in the fourth triangle above. It is an obtuse angle.

3 Find the area of the first triangle above.

Linear graphs

Linear graphs

D

- One method of drawing linear graphs is to **plot points**.

- You only need **two points** to draw a straight-line graph. However, it is better to use **three points**, because the the third point acts as a **check**.

- **Plot** the points you have found and **join** them up to draw the line.

Draw the graph of $y = 3x - 1$.

Pick a value for x, say $x = 3$, and work out the equivalent y-value.

$y = 3 \times 3 - 1 = 8$.

This gives the coordinates (3, 8).

Repeat this for another value of x, such as $x = 1$.

$y = 3 \times 1 - 1 = 2$ giving the coordinates (1, 2).

Choose a third value of x, such as $x = 0$.

$y = 3 \times 0 - 1 = -1$ giving the coordinates (0, -1).

Gradients

C

- The **gradient** of a line is a measure of its slope.

- It is calculated by **dividing** the **vertical** distance between two points on the line by the **horizontal** distance between the same two points.

$$\text{gradient} = \frac{\text{distance measured up}}{\text{distance measured along}}$$

- This is sometimes written as: $\text{gradient} = \dfrac{y\text{-step}}{x\text{-step}}$

These lines have gradients as shown.

- Lines that slope down from left to right have **negative** gradients.

- Note the **right-angled triangles** drawn along grid lines. These are used to find gradients.

- To draw a line with a certain **gradient**, for every unit moved **horizontally**, move upwards (or downwards if the gradient is negative) by the number of units of the gradient.

Draw lines with gradients of $\frac{1}{4}$ and -2.

Questions

Grade D

1 Draw a set of axes with x-values from -3 to $+3$ and y-values from -11 to $+13$. Draw the graph of $y = 4x + 1$ for values of x from -3 to $+3$.

Grade C

2 Find the gradient of the graph you drew in question **1**.

The gradient-intercept method

C

- The **gradient-intercept** method is the most straightforward and quickest method for drawing graphs.

- In the function $y = 2x + 3$, the **coefficient** of x (2) is the **gradient** and the **constant** term (+3) is the **intercept**.

- The **intercept** is the point where the line **crosses the y-axis**.

Draw the graph of $y = 2x + 3$.

First, mark the intercept point (0, 3).

Next, move 1 unit across and 2 units up to show the gradient. Repeat this for each new point.

Finally, join up the points to get the required line.

Cover-up method for drawing graphs

B

- This method is used to draw graphs of equations in the form $ax + by = c$.
- The x-axis has the equation $y = 0$. This means that all points on the x-axis have a y-value of 0.
- The y-axis has the equation $x = 0$. This means that all points on the y-axis have a x-value of 0.
- These facts can be used to draw any line that has an equation of the form $ax + by = c$.

Draw the graph of $2x + 3y = 12$.

Because the value of y is 0 on the x-axis, we can solve the equation for x: $2x + 3(0) = 12 \Rightarrow x = 6$

Similarly, because the value of x is 0 on the on the y-axis, we can also solve the equation for y: $2(0) + 3y = 12 \Rightarrow y = 4$

Hence, the line passes through the points (6, 0) and (0, 4).

Normally we would like a third point, but we can accept two because they are on the axes.

This type of equation can be drawn very easily using the cover-up method.

Start with the equation $2x + 3y = 12$

Cover up the x-term: $(2x) + 3y = 12 \Rightarrow$ when $x = 0$, $y = 4$

Cover up the y-term: $2x + (3y) = 12 \Rightarrow$ when $y = 0$, $x = 6$

This gives the points (6, 0) and (0, 4).

Questions

(Grade C)

1 Here are the equations of four lines.

A: $y = 2x - 3$ B: $y = 3x - 3$

C: $y = 2x + 1$ D: $y = \frac{1}{2}x - 1$

  a Which two lines are parallel?

  b Which two lines cross the y-axis at the same point?

(Grade C)

2 a Draw a set of axes with x-values from −3 to +3 and y-values from −9 to +15. On these axes, draw the graph of $y = 4x + 3$.

  b Draw a set of axes with x-values from −4 to +4 and y-values from −1 to +3. On these axes, draw the graph of $y = \frac{1}{2}x + 1$.

(Grade B)

3 Draw these lines using the cover-up method. Use the same grid, taking x from 0 to 6 and y from 0 to 6.

  a $3x + 4y = 12$ **b** $2x + 5y = 10$

Equations of lines

Finding the equation of a line from its graph

- When a graph is expressed in the form $y = mx + c$, the coefficient of x, m, is the **gradient**, and the constant term, c, is the **intercept** on the y-axis .

- This means that if we can find the gradient, m, of a line and its intercept, c, on the y-axis, we can write down the equation of the line.

Find the equation of the line shown.

First, find the y-intercept. This is at $(0, 1)$, so $c = 1$.

Next, measure the gradient of the line. For every move of 1 unit across, we move 2 units up which is a gradient of 2, so $m = 2$.

Hence the equation of the line is $y = 2x + 1$.

Uses of graphs – finding formulae or rules

- You revised real-life graphs on page 66. You can obtain formulae and equations from these graphs.

Top Tip!

Be careful with different scales on the axes when working out gradients.

This graph shows the cost of mobile phone calls on a monthly tariff.

The graph shows that there is a fixed charge of £10 per month. This is shown on the graph by the y-intercept (10). There is an additional charge of 20p per minute for all calls. This is shown by the gradient of the line, which is $\frac{1}{5}$.

Hence the equation for the cost of calls (in pounds) is $C = \frac{1}{5}m + 10$, where m is the number of minutes of calls.

So, for 60 minutes of calls the cost is $C = \frac{1}{5} \times 60 + 10 = £22$, which can be checked from the graph.

Questions

(Grade B)

1 What is the equation of the line shown?

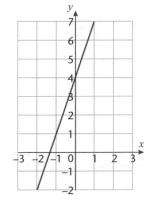

(Grade B)

2 This graph shows the cost of a taxi journey.

  a How much will a journey of 3 km cost?

  b What is the gradient of the line?

  c What is the intercept with the cost axis?

  d Write down an equation for the cost of a journey, C, in terms of the distance, d.

  e Work out the value of the equation in part **d** when $d = 3$.

Linear graphs and equations

Uses of graphs – solving simultaneous equations

- You can find the solution to two simultaneous equations by drawing their graphs on the same pair of axes.

By drawing their graphs on the same grid, find the solution of these simultaneous equations. $2x + y = 6$

$$y = 2x$$

First, draw the graph $2x + y = 6$ using the cover-up method.

It crosses the x-axis at (3, 0) and the y-axis at (0, 6).

Next, draw the graph of $y = 2x$ using the gradient-intercept method.

It passes through the origin and has a gradient of 2.

The graphs intersect at (1.5, 3). So the solution to the simultaneous equations is $x = 1.5$, $y = 3$.

Parallel and perpendicular lines

- If two lines are **parallel** their gradients are the same.

- If two lines are **perpendicular** their gradients are **negative reciprocals** of each other. For example, lines with a gradient of 2 and $-\frac{1}{2}$ are perpendicular to each other.

Top Tip!

Questions on perpendicular lines are usually accompanied by a diagram.

Find the equation of the line perpendicular to the midpoint of AB.

Find the midpoint of AB: M (2, 3)

Next, find the gradient of the line AB: 2 and work out the gradient of the perpendicular line: $-\frac{1}{2}$

Now, find the y-intercept. This can be done by drawing the line on the grid or by substituting the known point (2, 3) into $y = mx + c$.

$3 = -\frac{1}{2} \times 2 + c \Rightarrow 3 = -1 + c \Rightarrow c = 4$

Hence the equation of the required line is $y = -\frac{1}{2}x + 4$.

Questions

1 By drawing their graphs on the axes supplied, solve these simultaneous equations.

$y = 2x - 1$
$3x + 2y = 12$

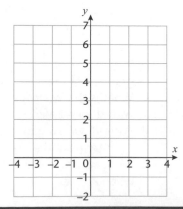

2 Find the equation of the line perpendicular to $y = \frac{1}{3}x + 2$ which passes through the point (0, 5).

Quadratic graphs

Drawing quadratic graphs

- A quadratic graph has an x^2 term in its equation.

  $y = x^2$ and $y = x^2 + 2x + 3$ will give quadratic graphs.

- Quadratic graphs always have the same characteristic shape, which is called a parabola.

- Quadratic graphs are drawn from tables of values.

  This table shows the values of $y = x^2 + 2x - 3$ for values of x from -4 to 2.

x	-4	-3	-2	-1	0	1	2
y	5	0	-3	-4	-3	0	5

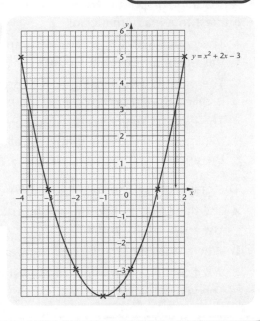

Top Tip!

As all parabolas have line symmetry, there will be some symmetry in the y-values in the table.

Top Tip!

Try to draw a smooth curve through all the points. Examiners prefer a good attempt at a curve rather than points joined with a ruler.

Reading values from quadratic graphs

- Once a quadratic graph is drawn it can be used to solve various equations.

  Use the graph of $y = x^2 + 2x - 3$ to find the x-values when $y = 3$.

  Draw the line $y = 3$ and draw down to the x-axis from the points where the line intercepts the curve, as shown above.

  The x-values are about $+1.7$ and -3.7.

Top Tip!

The graph is $y = x^2 - 3x - 4$ and the equation is $x^2 - 3x - 4 = 0$. The solution to the equation is found where the graph crosses the x-axis, where $y = 0$.

Using graphs to solve quadratic equations

- Solving a quadratic equation means finding the x-values that make it true.

- To solve a quadratic equation from its graph, read the values where the curve crosses the x-axis.

  Solve the equation $x^2 - 3x - 4 = 0$.

  The graph crosses the x-axis at $x = -1$ and $x = 4$.

Questions

Grade C

1 a Set up and complete the table of values for $y = x^2 + x - 1$ for $-3 \leqslant x \leqslant 3$.

 b Draw the graph of $y = x^2 + x - 1$. Label the x-axis from 3 to $+3$ and the y-axis from -2 to 12.

 c Use the graph to find the values of x when $y = 2$.

 d Use the graph to solve the equation $x^2 + x - 1 = 0$.

Non-linear graphs

The significant points of a quadratic graph

A*

- A quadratic graph has **four points** that are of interest to a mathematician.

  - The points A and B, where the graph crosses the x-axis, are called the **roots**

  - The point C, where the graph crosses the y-axis, is called the **intercept**

  - The point D, which is the maximum or minimum point of the graph, is called the **vertex**

The graph of $y = x^2 - 3x - 4$ is shown.

a Use the graph to find the solutions to $x^2 - 3x - 4 = 0$.
The roots are –1 and 4.

b Where does the graph cross the y-axis? The intercept is –4.

c What are the coordinates of the vertex?
The vertex is at (1.5, –6.25).

Solving equations by the method of intersection

A*

- Many equations can be solved by drawing two intersecting graphs on the same axes and using the x-value(s) of their point(s) of intersection.

- This usually involves of the intersection of a curve (usually a quadratic) and a straight line that is used to solve a new equation.

- The method for finding the straight line has four steps.

Given the graph of $y = x^2 + 3x - 2$, draw a suitable straight line to solve the equation $x^2 + 2x - 3 = 0$.

- Step 1: Write down the original (given) equation $\quad\quad y = x^2 + 3x - 2$
- Step 2: Write down the (new) equation to be solved in reverse $\quad 0 = x^2 + 2x - 3$
- Step 3: Subtract these equations $\quad\quad\quad\quad\quad\quad\quad\quad y = x + 1$
- Step 4: Draw the line given by the answer

Questions

Grades A-A*

1 The graph of $y = x^2 + 3x - 2$ is shown opposite.

a Write down the coordinates of the intercept with the y-axis.

b Write down the coordinates of the vertex.

c What are the roots of the equation $x^2 + 2x - 3 = 0$?

d By drawing a suitable straight line, solve the equation $x^2 + 3x - 1 = 0$.

e By drawing another straight line, solve the equation $x^2 + 2x - 2 = 0$.

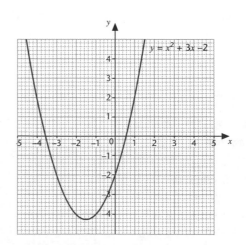

Other graphs

Reciprocal graphs

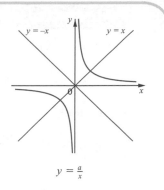

- A reciprocal equation has the form $y = \dfrac{a}{x}$.
- All reciprocal graphs have a similar shape and some symmetry properties.
  - The lines $y = x$ and $y = -x$ are lines of symmetry.
  - The closer x gets to zero, the nearer the graph gets to the y-axis.
  - As x increases, the graph gets closer to the x-axis.
  - The graph never actually touches the axes, but gets closer and closer to them. The axes are called **asymptotes**.

Cubic graphs

- A cubic function or graph is one that contains a term in x^3.
- You should be able to recognise the characteristic shape of the basic function $y = x^3$.
- You should be able to draw other cubic graphs from a table of values.

This partly completed table shows $y = x^3 - x$ for values of x from -3 to 3.

x	-3	-2.5	-2	-1.5	-1	-0.5	0	0.5	1	1.5	2	2.5	3
y	-24	-13.13	-6	-1.88				-0.38	0	1.88	6		

Exponential graphs

- Equations that have the form $y = k^x$, where k is a positive number, are called **exponential functions**.
- Exponential functions share the following properties.
  - When k is greater than 1, the value of y increases steeply as x increases
  - Also, when k is greater than 1, as x takes on increasing large negative values, the closer y gets to zero, and so the negative x-axis becomes an asymptote
  - Whatever the value of k, the graph always crosses the y-axis at 1
  - The reciprocal graph, $y = k^{-x}$, is the reflection in the y-axis of $y = k^x$
  - When k is less than 1, the graph is the same shape as $y = k^{-x}$
- Exponential graphs are always drawn from a table of values.

This partly completed table shows $y = 2^x$ for values of x from 0 to 4.

x	0	0.5	1	1.5	2	2.5	3	3.5	4
y	1	1.41	2	2.83	4				16

Questions

1 Complete the table of values for $y = x^3 - x$ above and use it to draw the graph of $y = x^3 - x$ for values of x from -3 to $+3$. (Take the y-axis from -24 to $+24$.)

2 Complete the table for $y = 2^x$ above and use it to draw the graph of $y = 2^x$ for values of x from 0 to 4. (Take the y-axis from 0 to $+16$.)

Algebraic fractions

Algebraic fractions

- The following four rules are used to work out the value of fractions.

Addition: $\dfrac{a}{b} + \dfrac{c}{d} \equiv \dfrac{ad + bc}{bd}$

Subtraction: $\dfrac{a}{b} - \dfrac{c}{d} \equiv \dfrac{ad - bc}{bd}$

Multiplication: $\dfrac{a}{b} \times \dfrac{c}{d} \equiv \dfrac{ac}{bd}$

Division: $\dfrac{a}{b} \div \dfrac{c}{d} \equiv \dfrac{ad}{bc}$

- Note that a, b, c and d can be numbers, letters or algebraic expressions.

- The following points make working with expressions easier.

  - Use **brackets**

  - **Factorise** if you can

  - **Cancel** if you can

> **Top Tip!**
>
> Use brackets around terms in the rules before you work them out. This makes it less likely that you will make a mistake.

Simplify $\dfrac{2y}{x} + \dfrac{x}{2y}$.

Using the addition rule: $\dfrac{(2y)(2y) + (x)(x)}{(x)(2y)} = \dfrac{4y^2 + x^2}{2xy}$

Simplify $\dfrac{2}{x} - \dfrac{y}{3x}$.

Using the subtraction rule: $\dfrac{(2)(3x) - (x)(y)}{(x)(3x)} = \dfrac{6x - xy}{3x^2} = \dfrac{x(6 - y)}{3x^2} = \dfrac{6 - y}{3x}$

Simplify $\dfrac{a}{2} \times \dfrac{a + b}{a - b}$.

Using the multiplication rule: $\dfrac{(a)(a + b)}{(2)(a - b)} = \dfrac{a^2 + ab}{2a - 2b}$

Simplify $\dfrac{a}{4} \div \dfrac{3a}{8}$.

Using the division rule: $\dfrac{(a)(8)^2}{(4)^1(3a)} = \dfrac{2}{3}$

> **Top Tip!**
>
> You will be expected to simplify answers as much as possible, so remember to look out for common factors.

Questions

Grade A

1 Simplify each of these.

a $\dfrac{x + 1}{2} + \dfrac{x - 2}{3}$ **b** $\dfrac{x}{5} - \dfrac{2x + 1}{3}$ **c** $\dfrac{2x}{5} \times \dfrac{3y}{4}$ **d** $\dfrac{4x}{3y} \div \dfrac{x}{y}$

Solving equations

Solving equations with algebraic fractions

- The algebraic fraction identities on page 80 can be used to solve equations.

Solve $\dfrac{x+1}{3} - \dfrac{x-3}{2} = 1$.

Use the subtraction rule for combining fractions, and also cross-multiply the denominator of the left-hand side to the right-hand side.

$\dfrac{(x+1)(2) - (3)(x-3)}{(3)(2)} = 1 \Rightarrow 2(x+1) - 3(x-3) = (1)(3)(2)$

$2x + 2 - 3x + 9 = 6 \Rightarrow -x = -5 \Rightarrow x = 5$

> **Top Tip!**
> Use brackets around all numbers and expressions. Most marks are lost in this type of problem because of errors with minus signs.

Solve $\dfrac{3}{x-1} - \dfrac{2}{x+1} = 1$.

Use the subtraction rule for combining fractions, and cross-multiply the denominator, using the method shown in the previous example.

$(3)(x+1) - (2)(x-1) = (x-1)(x+1)$

$3x + 3 - 2x + 2 = x^2 - 1$

(Right-hand side is the difference of two squares.)

Rearrange into the general quadratic form.

$x^2 - x - 6 = 0$

Factorise and solve the equation.

$(x-3)(x+2) = 0 \Rightarrow x = 3$ or $x = -2$

> **Top Tip!**
> When you rearrange your equation into the quadratic it should factorise. If it doesn't, then you will almost certainly have made a mistake unless the question states that it requires an answer as a surd or a decimal.

Simplify $\dfrac{2x^2 + x - 3}{4x^2 - 9}$.

Factorise the numerator and denominator:

$\dfrac{(2x+3)(x-1)}{(2x+3)(2x-3)}$

Cancel any common factors:

$\dfrac{x-1}{2x-3}$

> **Top Tip!**
> There will always be a common factor on the top and the bottom which will cancel out.

Questions

Grade A*

1 Solve these equations.

a $\dfrac{x+1}{2} + \dfrac{x-2}{3} = 4$

b $\dfrac{4}{x+1} + \dfrac{5}{x+2} = 2$

Grade A*

2 Simplify these expressions.

a $\dfrac{2x^2 + x - 1}{4x^2 - 1}$

b $\dfrac{2x^2 + x - 3}{4x^2 - 9}$

Simultaneous equations 2

Linear and non-linear simultaneous equations

- On page 59, you revised solving two linear simultaneous equations by the method of substitution.

- You can use a similar method to solve a pair of a equations, one of which is linear and the other of which is non-linear, but you must always substitute the linear equation into the non-linear equation. This usually gives a soluble quadratic equation.

Solve these simultaneous equations.

$$x^2 + y^2 = 5 \ (1)$$
$$x + y = 3 \quad (2)$$

Rearrange equation (2) to obtain:

$$y = 3 - x$$

Substitute this into the non-linear equation (1):

$$x^2 + (3 - x)^2 = 5$$

Expand: $\qquad x^2 + 9 - 6x + x^2 = 5$

Rearrange into the general quadratic form:

$$2x^2 - 6x + 4 = 0$$

Cancel by 2: $\qquad x^2 - 3x + 2 = 0$

Factorise and solve: $\qquad (x - 1)(x - 2) = 0 \Rightarrow x = 1 \text{ or } x = 2$

Substitute x into equation (2) to find values of y:

$$\text{when } x = 1, y = 2$$
$$\text{and when } x = 2, y = 1$$

So the solutions are (1, 2) and (2, 1).

Top Tip!

Remember to give your answers as pairs of values in x and y. Marks are often lost in examinations by not giving a full solution.

Solve these simultaneous equations.

$$y = x^2 - 2x \ (1)$$
$$y = 2x - 3 \ (2)$$

In this case, the equations can just be put equal to each other so:

$$x^2 - 2x = 2x - 3$$

Rearrange into the general quadratic form:

$$x^2 - 4x + 3 = 0$$

Factorise and solve: $\qquad (x - 1)(x - 3) = 0 \Rightarrow x = 1 \text{ or } x = 3$

Substitute x into equation (2) to find values of y:

$$\text{when } x = 1, y = -1 \text{ and when } x = 3, y = 3$$

So the solutions are (1, −1) and (3, 3).

Top Tip!

When you rearrange into a quadratic, it will factorise. If it doesn't you will have made a mistake.

Questions

Grade A*

1 Solve these pairs of simultaneous equations.

   a $x^2 + y^2 = 25$ **b** $x^2 + y^2 = 13$
      $x - y = 1$ $x = 13 - 5y$

The nth term

nth term of a sequence

- A linear sequence has the same difference between consecutive terms.

  3, 8, 13, 18, 23, 28, ... is a linear sequence with a constant difference of 5.

- The nth term of a linear sequence is always of the form $An \pm b$.

  $3n + 1$, $4n - 3$ and $8n + 7$ are examples of nth terms of a linear sequence.

- To find the coefficient of n, A, find the difference between consecutive terms.

  The sequence 4, 7, 10, 13, 16, ... has a constant difference of 3, so the nth term of the sequence will be given by $3n \pm b$.

- To find the value of b, work out the difference between A and the first term of the sequence.

  The sequence 4, 7, 10, 13, 16, ... has a first term of 4. The coefficient of n is 3. To get from 3 to 4 you add 1, so the nth term is $3n + 1$.

  Find the nth term of the sequence 4, 9, 14, 19, 24, ...
  The constant difference is 5, and $5 - 1 = 4$, so the nth term is $5n - 1$.

Special sequences

- There are many **special sequences** that you should be able to recognise.
  - The even numbers   2, 4, 6, 8, 10, 12, ... The nth term is $2n$.
  - The odd numbers   1, 3, 5, 7, 9, 11, ... The nth term is $2n - 1$.
  - The square numbers   1, 4, 9, 16, 25, 36, ... The nth term is n^2.
  - The triangular numbers   1, 3, 6, 10, 15, 21, ... The nth term is $\frac{1}{2}n(n + 1)$.
  - The powers of 2   2, 4, 8, 16, 32, 64, ... The nth term is 2^n.
  - The powers of 10   10, 100, 1000, 10 000, 100 000, ... The nth term is 10^n.
  - The prime numbers   2, 3, 5, 7, 11, 13, 17, 19, ... There is no nth term as there is no pattern to the prime numbers.

Top Tip!
The only even prime number is 2.

Questions

Grade C

1 a The nth term of a sequence is $5n - 1$.
   i Write down the first three terms of the sequence.
   ii Write down the 100th term of the sequence.
  b The nth term of a sequence is $\frac{1}{2}(n + 1)(n + 2)$.
   i Write down the first three terms of the sequence.
   ii Write down the 199th term of the sequence.

Grade C

2 Write down the nth term of each of these sequences.
  a 6, 11, 16, 21, 26, 31, ...
  b 3, 11, 19, 27, 35, ...
  c 9, 12, 15, 18, 21, 24, ...

Grade C

3 a What is the 100th even number?
  b What is the 100th odd number?
  c What is the 100th square number?
  d Continue the sequence of triangular numbers up to 10 terms.
  e Continue the sequence of powers of 2 up to 10 terms.
  f Write down the next five prime numbers after 19.

Formulae

The nth term from given patterns

- An important part of mathematics is to find **patterns in situations**.

- Once a **pattern** has been found, the **nth term** can be used to describe the pattern.

  The diagram shows a series of 'L' shapes.

   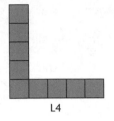

L1   L2   L3   L4

> **Top Tip!**
> Write down the sequence then look for the nth term.

How many squares are there in L100?

First write down the sequence of squares in the 'L' shapes: 3, 5, 7, 9, ...

The constant difference is 2, and $2 + 1 = 3$, so the nth term is $2n + 1$.

The 100th term will be $2 \times 100 + 1 = 201$, so there will be 201 squares in L100.

Changing the subject of a formula

- On page 60, you revised changing the **subject** of a formula in which the subject appears once.

- The principle is the same when the the subject appears more than once.

Make x the subject of $2x + y = ax + b$.

First, rearrange the formula to get all the x terms on the left-hand side and the other terms on the right-hand side: $2x - ax = b - y$

Take a factor of x out of the left-hand side to get: $x(2 - a) = b - y$

Divide by the bracket, which gives: $x = \dfrac{b - y}{2 - a}$

Make x the subject of $y = \dfrac{x - 2}{x + 3}$.

First, cross-multiply which gives: $y(x + 3) = x - 2$

Expand the bracket to get: $yx + 3y = x - 2$

Collect subject terms: $yx - x = -2 - 3y$

Take a factor of x out of the left-hand side to get: $x(y - 1) = -2 - 3y$

Divide by the bracket, which gives: $x = \dfrac{-2 - 3y}{y - 1}$

> **Top Tip!**
> When the subject appears twice, you will need to factorise and divide by the bracket at some point in your working.

Questions

1 Matches are used to make pentagonal patterns.

1   2   3   4

a How many matches will be needed to make the 10th pattern?

b How many matches will be needed to make the nth pattern?

2 Make x the subject of each of these formulae.

   a $ax + 6y = 2x + 8y$ **b** $y = \dfrac{x + 4}{x - 2}$

Inequalities

Solving inequalities

- An **inequality** is an algebraic expression that uses the signs $<$ (less than), $>$ (greater than), $\leqslant$ (less than or equal to) and $\geqslant$ (greater than or equal to).

- The solutions to inequalities are a **range of values**.

  The expression $x < 2$ means that x can take any value less than 2, all the way to minus infinity. x can also be very close to 2, for example 1.9999..., but is never actually 2.

  $x \geqslant 3$ means x can take any value greater than 3, up to infinity, or 3 itself.

- **Linear inequalities** can be solved using the same rules that you use to solve equations.

- The answer when a linear inequality is solved is an inequality such as $x > -1$.

  Solve $\frac{x}{5} - 3 \geqslant 4$.

  Add 3 to both sides: $\qquad \frac{x}{5} \geqslant 7$

  Multiply both sides by 5: $\qquad x \geqslant 35$

Top Tip!

Don't use the equals signs when solving inequalities as this doesn't get any marks in an examination.

Inequalites on number lines

- The solution to a linear inequality can be shown on a **number line** by using the convention that an open circle is a **strict inequality** and a filled-in circle is an **inclusive inequality**.

represents $x < 3$.   represents $x > 1$.   represents $x \leqslant -2$.

represents $x \geqslant 4$.   represents $-1 \leqslant x < 2$.

Top Tip!

Inequalities often ask for integer values. An integer is a positive or negative whole number, including zero.

Solve the inequality $2x + 3 < 11$ and show the solution on a number line.

The solution is $x < 4$, which is shown on this number line.

Questions

Grade C

1 Solve the following inequalities.

  a $x + 5 < 8$ **b** $2x + 3 > 5$

  c $\frac{x}{3} - 5 \geqslant 1$ **d** $\frac{x}{2} + 7 > 2$

Grade C

2 What inequalities are shown by the following number lines?

  a

  b

Grade C

3 a What inequality is shown on this number line?

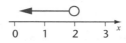

  b Solve the inequality $3x + 6 \geqslant 3$.

  c What integers satisfy both of the inequalities in parts **a** and **b**?

Graphical inequalities

Graphical inequalities

- A linear **inequality** can be plotted on a graph.

- The result is a **region** that lies on one side or the other of a straight **boundary line**.

  $y < 2$, $x > -3$, $y \geqslant 2x + 1$ and $2x + 3y < 6$ are examples of linear inequalities which can be represented on a graph.

- The first step is to draw the boundary line. This is found by replacing the inequality sign with an equals sign.
  - If the inequality is strict ($<$ or $>$), the boundary line should be dashed
  - If the inequality is inclusive ($\leqslant$ or $\geqslant$), the boundary line should be solid

- Once the boundary line is drawn, the required region is shaded.
  - To confirm on which side of the line the region lies, choose any point that is not on the boundary line and test it in the inequality. If it satisfies the inequality, that is the side required. If it doesn't, the other side is required.

Top Tip!

Use the origin as the point to test if possible as it makes working out the inequality easier.

Show both of these inequalities on a graph.

a $y \leqslant 3$ **b** $y < 2x - 1$

First draw the boundary lines:

a $y = 3$ (solid)   **b** $y = 2x - 1$ (dashed)

Test a point that is not on the line, such as (0, 0).

More than one inequality

- You will be required to show more than one inequality on the same graph.

- In this case, it is clearer to shade the regions that are *not* required, so that the required region is left unshaded.

On the same grid, show the region bounded by these inequalities.

$x \geqslant 1$ $x + y < 5$ $y > x$

Start by drawing the boundary lines:

$x = 1$ (solid)
$x + y = 5$ (dashed)
$y = x$ (dashed)

Test a point that is not on the line in each case and shade the appropriate regions.

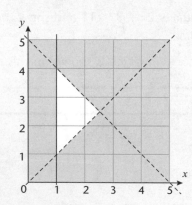

Questions

Grade B

1 On separate grids, with x-and y-values from 0 to 5, show the following inequalities.

   a $x < 4$ **b** $x + y \leqslant 2$ **c** $4y \geqslant 5x - 1$

Grade B

2 On the same grid with x-and y-values from 0 to 5, show the region bounded by these inequalities.

   $y \leqslant 3$ $x + y \geqslant 3$ $y \geqslant x$

Graph transforms

A*

Transformations of the graph $y = f(x)$

- The notation $f(x)$ is used to represent a **function** of x. A function of x is any algebraic expresion in which x is the only variable.

  $f(x) = x + 5$, $f(x) = 7x$, $f(x) = 3x + 2$ and $f(x) = x^2$ are all functions of x.

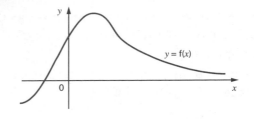

- The graph $y = f(x)$ can be transformed by six rules that you should understand. Three are covered in this section and three are covered in the next section.

  - **Rule 1** The graph of $y = f(x) + a$ is a translation of the graph $y = f(x)$ by the vector $\binom{0}{a}$

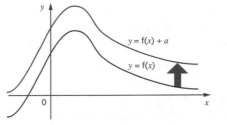

  - **Rule 2** The graph of $y = f(x - a)$ is a translation of the graph $y = f(x)$ by the vector $\binom{a}{0}$

Top Tip!

The vector for $f(x - a)$ has the opposite sign to the number in the bracket.

  - **Rule 3** The graph of $y = kf(x)$ is a stretch of the graph $y = f(x)$ by a scale factor of k in the y-direction

The graph of $y = \sin x$ is shown by a dotted line on each of the grids below for $0° \leqslant x \leqslant 360°$.

What are the equations of the graphs shown by the solid red lines?

a is a translation upwards by 1 unit so the graph is $y = \sin x + 1$.

b is a stretch of scale factor 2 in the y-direction so the graph is $y = 2 \sin x$.

Questions

Grade A*

1 Draw a sketch of $y = \cos x$ for $0° \leqslant x \leqslant 360°$. On the same axes, sketch the graph of $y = \cos x - 1$.

Grade A*

2 Draw a sketch of $y = \cos x$ for $0° \leqslant x \leqslant 360°$. On the same axes, sketch the graph of $y = \frac{1}{2} \cos x$.

Grade A*

3 Draw a sketch of $y = x^2$. Label both axes -5 to $+5$. On the same axes, sketch the graph of $y = (x - 1)^2$.

Remember: You must revise all content from Grade E to the level that you are currently working at.

Transformations of the graph $y = f(x)$

A*

- The graph $y = f(x)$ can be transformed by six rules that you should understand. Rules 4 to 6 are covered here.

  - **Rule 4** The graph of $y = f(tx)$ is a stretch of the graph $y = f(x)$ by a scale factor of $\frac{1}{t}$ in the x-direction.

  - **Rule 5** The graph of $y = -f(x)$ is a reflection of the graph $y = f(x)$ in the x-axis.

  - **Rule 6** The graph of $y = f(-x)$ is a reflection of the graph $y = f(x)$ in the y-axis.

Top Tip!

The scale factor of the stretch is the reciprocal of the number inside the bracket.

The graph of $y = \sin x$ is shown by a dotted line on each of the grids below for $0° \leqslant x \leqslant 360°$.

What are the equations of the graphs shown by the solid red lines?

a

b

a is a stretch by scale factor $\frac{1}{2}$ in the x-direction, so the graph is $y = \sin 2x$.

b is a reflection in the x-axis, so the graph is $y = -\sin x$.

The graph of $y = x^3 + 1$ is shown by a dotted line on the axes on the right.

On the same axes, sketch the graph of $y = -x^3 + 1$.

The graph is $y = (-x)^3 + 1$ which is a reflection of the curve in the y-axis. This is the graph shown by a solid line.

Questions

1 Draw a sketch of $y = \cos x$ for $0° \leqslant x \leqslant 360°$. On the same axes, sketch the graph of $y = \cos 3x$.

2 Draw a sketch of $y = \cos x$ for $0° \leqslant x \leqslant 360°$. On the same axes, sketch the graph of $y = -\cos x$.

3 Draw a sketch of $y = x^3 + 1$. Label the x axis from -5 to $+5$ and the y axis from -10 to $+10$. On the same axes, sketch the graph of $y = -x^3 - 1$.

Proof

Proof

- The method of mathematical **proof** is to proceed in logical steps, establishing a series of mathematical statements by using facts that are already known to be true.

- Each step must follow from the previous step and all statements must be given reasons if necessary.

Geometric proof

- Geometric proofs are usually concerned with circle theorems or congruency.

- They use the four conditions for congruency – SSS, SAS, ASA and RHS – which you revised on page 42.

- In a geometric proof you must give reasons for the values you find for the angles.

ABCD is a parallelogram. X is the point where the diagonals meet. Prove that triangles AXB and CXD are congruent.

$\angle$BAX = $\angle$DCX (alternate angles)

$\angle$ABX = $\angle$CDX (alternate angles)

AB = CD (opposite sides in a parallelogram)

Hence $\triangle$AXB is congruent to $\triangle$CXD (ASA).

Algebraic proof

- Algebraic proofs are usually concerned with proving that two algebraic expressions are equivalent.

- You usually start with the left-hand expressions and use the standard algebraic techniques of expansion, simplification, factorisation and cancelling to prove that this is equal to the right-hand side.

- You must make sure that all the algebra is well explained and each step is clear. Do not assume that an examiner will fill in the gaps for you.

Prove that $(n + 1)^2 + n^2 - (n - 1)^2 = n(n + 4)$.

Expand the LHS to get: $n^2 + 2n + 1 + n^2 - (n^2 - 2n + 1)$

Show clearly that the minus sign in front of the second bracket is properly dealt with:

$$n^2 + 2n + 1 + n^2 - n^2 + 2n - 1$$

Collect like terms: $n^2 + 4n$

Factorise the collected result: $n^2 + 4n = n(n + 4)$ which is the RHS of the original expression.

Questions

(Grade A*)

1 Prove that the exterior angle of a triangle, z, is equal to the sum of the two opposite angles, x and y.

(Grade A*)

2 Prove that $(n + 6)^2 - (n + 2)^2 = 8(n + 4)$.

Algebra checklist

I can...

- [ ] use letters to write more complicated algebraic expressions
- [ ] expand expressions with brackets
- [ ] factorise simple expressions
- [ ] solve linear equations where the variable appears on both sides of the equals sign
- [ ] solve linear equations that require the expansion of a bracket
- [ ] set up and solve simple equations from real-life situations
- [ ] find the average speed from a travel graph
- [ ] draw a linear graph without a table of values
- [ ] substitute numbers into an nth-term rule
- [ ] understand how odd and even numbers interact in addition, subtraction and multiplication problems

You are working at (Grade D) level.

- [ ] expand and simplify expressions involving brackets
- [ ] factorise expressions involving letters and numbers
- [ ] expand pairs of linear brackets to give a quadratic expression
- [ ] solve linear equations that have the variable on both sides and include brackets
- [ ] solve simple linear inequalities
- [ ] show inequalities on a number line
- [ ] solve equations, using trial and improvement
- [ ] rearrange simple formulae
- [ ] give the nth term of a linear sequence
- [ ] give the nth term of a sequence of powers of 2 or 10
- [ ] draw a quadratic graph, using a table of values

You are working at (Grade C) level.

- [ ] solve a quadratic equation from a graph
- [ ] recognise the shape of graphs $y = x^3$ and $y = \dfrac{1}{x}$
- [ ] solve two linear simultaneous equations
- [ ] rearrange more complex formula to make another variable the subject
- [ ] factorise a quadratic expression of the form $x^2 + ax + b$
- [ ] solve a quadratic equation of the form $x^2 + ax + b = 0$
- [ ] interpret real-life graphs
- [ ] find the equation of a given linear graph
- [ ] solve a pair of linear simultaneous equations from their graphs
- [ ] draw cubic graphs, using a table of values
- [ ] use the nth term to generate a quadratic sequence

- [ ] solve equations involving algebraic fractions where the subject appears as the numerator
- [ ] solve more complex linear inequalities
- [ ] represent a graphical inequality on a coordinate grid
- [ ] find the inequality represented by a graphical inequality
- [ ] verify results by substituting numbers

You are working at (Grade B) level.

- [ ] draw exponential and reciprocal graphs, using a table of values
- [ ] find the proportionality equation from a direct or inverse proportion problem
- [ ] set up and solve two simultaneous equations from a practical problem
- [ ] factorise a quadratic expression of the form $ax^2 + bx + c$
- [ ] solve a quadratic equation of the form $ax^2 + bx + c = 0$ by factorisation
- [ ] solve a quadratic equation of the form $ax^2 + bx + c = 0$ by the quadratic formula
- [ ] write a quadratic expression of the form $x^2 + ax + b$ in the form $(x + p)^2 + q$
- [ ] interpret and draw more complex real-life graphs
- [ ] find the equations of graphs parallel and perpendicular to other lines and passing through specific points
- [ ] rearrange a formula where the subject appears twice
- [ ] combine algebraic fractions, using the four rules of addition, subtraction, multiplication and division
- [ ] translate and solve a real-life problem, using inequalities
- [ ] show that a statement is true, using verbal or mathematical arguments

You are working at (Grade A) level.

- [ ] solve equations using the intersection of two graphs
- [ ] use trigonometric graphs to solve sine and cosine problems
- [ ] solve direct and inverse proportion problems, using three variables
- [ ] solve a quadratic equation of the form $x^2 + ax + b = 0$ by completing the square
- [ ] solve real-life problems that lead to a quadratic equation
- [ ] solve quadratic equations involving algebraic fractions where the subject appears as the denominator
- [ ] rearrange more complicated formula where the subject may appear twice
- [ ] simplify algebraic fractions by factorisation and cancellation
- [ ] solve a pair of simultaneous equations where one is linear and one is non-linear
- [ ] transform the graph of a given function
- [ ] identify the equation of a transformed graph
- [ ] prove algebraic results with rigorous and logical arguments

You are working at (Grade A*) level.

Answers

Page 4

1 a mode = 8, median = 7.5,
mean = 6.6

b mode = 11, median = 12,
mean = 12

2 a 30 **b** 0 **c** 1

d i 45 **ii** 1.5

Page 5

1 a $20 < x \leqslant 30$

b i 1350 **ii** 27

2 a 10

b Because there are 25 snails below 15
and 75 above 15

c 27.5 grams

Page 6

1 a about 30

b Because this is the next
4-point moving average.

c about 25

2 a One of: leading question, two
questions in one, double negative,
not enough responses

b One of: overlapping responses,
missing responses

Page 7

1 85

2 a i Not representative as only year 7

ii Could be random but students
who walk are excluded

iii This will give a random sample as
all students have an equal
chance.

b Y7: 12, Y8: 10, Y9: 12,
Y10: 11, Y11: 9

Page 8

1 a Town B – on average about 4
degrees hotter

b Town B – December is in summer

c September

d No – the lines cross but they have no
meaning (just show trends)

2 a 20 **b** 36 (49 – 13) **c** 35

d 26.5 **e** 28 (560 ÷ 20)

Page 9

1 a positive

b The higher the spelling marks, the
higher the tables marks

2 a As the car gets older, its value
decreases

b There is no relationship between the
distance someone lives from work
and the wages they earn

3 73

Page 10

1 a 66 **b** 48 **c** 83

d 36

e

Page 11

1 a Ali: 0.25 Barry: 0.22 Clarrie: 0.19

b Clarrie – most trials **c** 40

2 a i $\frac{1}{2}$ **ii** $\frac{1}{2}$

iii They cannot happen at same
time.

iv The probabilities add up to 1

b i $\frac{1}{13}$ **ii** $\frac{12}{13}$

Page 12

1 a 50 **b** 20 **c** 70

2 a 200 **b** 5% **c** 40%

3 a $\frac{4}{52} = \frac{1}{13}$ **b** $\frac{4}{52} = \frac{1}{13}$ **c** $\frac{8}{52} = \frac{2}{13}$

d $\frac{48}{52} = \frac{12}{13}$ **e** $\frac{44}{52} = \frac{11}{13}$

Page 13

1 a 36

b i $\frac{6}{36} = \frac{1}{6}$ **ii** $\frac{4}{36} = \frac{1}{9}$

c i $\frac{4}{36} = \frac{1}{9}$ **ii** $\frac{6}{36} = \frac{1}{6}$

iii 7

2 a $\frac{5}{12}$ **b** $\frac{1}{12}$ **c** $\frac{5}{12}$

Page 14

1 a $\frac{1}{16}$ **b** $\frac{7}{16}$

2 a $\frac{1}{15}$ **b** $\frac{2}{5}$ **c** $\frac{8}{15}$

Page 16

1 a 35 **b** £11.34

2 a 18 **b** 13 **c** 12

Page 17

1 a i 4 **ii** 0.8 **iii** 60

b i 57 **ii** 0.97 **iii** 89

2 a 700 **b** 5 **c** 10 **d** 32

3 a 300 **b** 760 **c** 0.075

d 0.034 **e** 600 000 **f** 200

Page 18

1 a i 30 **ii** 84 **iii** 130 **iv** 36

v 40 **vi** 300

b i $2^2 \times 5$ **ii** $3^2 \times 5$

iii 2^6 **iv** $2^3 \times 3 \times 5$

2 a i 30 **ii** 21 **iii** 39

b The LCM is the product of the two
numbers

c i 18 **ii** 40 **iii** 75

d i 6 **ii** 2 **iii** 5 **iv** 16 **v** 12

vi 1

Page 19

1 a $\frac{2}{5}$ **b** $\frac{3}{4}$ **c** $\frac{1}{3}$

2 a i $\frac{19}{28}$ **ii** $1\frac{5}{8}$ **iii** $6\frac{1}{15}$

b i $\frac{13}{30}$ **ii** $\frac{2}{9}$ **iii** $\frac{7}{12}$

c i $\frac{1}{6}$ **ii** $\frac{5}{14}$ **iii** $3\frac{17}{20}$

d i $\frac{7}{10}$ **ii** $1\frac{3}{4}$ **iii** $1\frac{3}{5}$

Page 20

1 a i 0.8 **ii** 0.07 **iii** 0.22

b i 1.05 **ii** 1.12 **iii** 1.032

c i 0.92 **ii** 0.85 **iii** 0.96

2 a £168 **b** 66.24 kg

3 a i 740 **ii** 5% **b** 20%

Page 21

1 a £614.63 **b** 1371

2 a £350 **b** 14 000

Page 22

1 a £100 and £400 **b** 50 g and 250 g

c £150 and £250 **d** 80 kg and 160 kg

2 a 56 **b** 30

3 a 37.5 mph **b** 52.5 km

Page 23

1 a 31.25 kg **b** 160

2 a travel-size (75 g) – 1.44 g/p
compared to 1.38 g/p for large tube
(250 g)

b 95 out of 120 – 79.2% compared to
62 out of 80 which is 77.5%

3 a 6.25 g/cm³ **b** 27 kg

c 0.3 m³

Page 24

1 a i 9 **ii** 4 **iii** 5

b i ±2 **ii** +1 **iii** −2

c i 24 **ii** 1.1 **iii** 6.1

2 a i 27 **ii** 64 **iii** 1000

b i 4^5 **ii** 6^6 **iii** 10^4 **iv** 2^7

c i 1024 **ii** 46 656

iii 10 000 **iv** 128

d 64, 128, 256, 512, 1024

3 a i $\frac{1}{64}$ **ii** $\frac{1}{7}$ **iii** $\frac{1}{9}$

b i 2^{-3} **ii** 3^{-1} **iii** x^{-n}

Page 25

1 **a i** 5.6×10^5 **ii** 7×10^{-6}
  iii 3×10^6
  b i 6 400 000 **ii** 0.000 83
  iii 900 000 000
  c i 1.56×10^7 **ii** 4×10^3
  iii 9×10^6 **iv** 8×10^{-5}
2 **a** $0.3\dot{6}$ **b** $0.6\dot{1}\dot{5}$ **c** $0.3\dot{6}$
  d $0.\dot{6}$ **e** $0.1\dot{6}$ **f** $0.\dot{7}$
3 **a i** 0.25 **ii** 0.05 **iii** $0.\dot{1}$
  b i $1\frac{1}{7}$ **ii** $1\frac{4}{5}$ **iii** $4\frac{1}{3}$

Page 26

1 **a i** 2^7 **ii** 2^9 **iii** x^9 **iv** 3^3
  v 3^4 **vi** x^4
  b i $4a^2b^2c^3$ **ii** $4x^2y^3z$
2 **a** x^{15} **b** $27a^6$ **c** $4x^2y^4$
3 **a** 5 **b** 4 **c** $\frac{1}{5}$
4 **a** $\frac{1}{16}$ **b** $\frac{1}{9}$ **c** $\frac{1}{100\,000}$

Page 27

1 **a** $\sqrt{4}$ **b** $6\sqrt{36}$ **c** $5\sqrt{3}$
  d $9\sqrt{2}$ **e** $3 + 2\sqrt{3}$
  f -1 **g** $14 - 6\sqrt{5}$
2 $7 + 3\sqrt{3}$
3 **a** $\frac{\sqrt{3}}{2}$ **b** $\frac{3}{2}$ **c** $\frac{3\sqrt{5} - 5}{5}$

Page 28

1 **a** $y = kx^2$ **b** $\frac{1}{20}$ or 0.05
  c 1.25 **d** 20
2 **a** $y = \frac{k}{\sqrt{x}}$ **b** 12 **c** 4 **d** 1

Page 29

1 **a** 6.5 m, 7.5 m
  b 33.5 kg, 34.5 kg
  c 315 cm, 325 cm
2 **a** $86.25 \leqslant$ area < 106.25
  b 29.1 mph $\leqslant$ speed < 35.5 mph

Page 31

1 **a** 37.7 cm **b** 8π cm
2 **a** 706.9 cm² **b** 9π cm²
3 **a** 37.5 cm² **b** 34 cm²

Page 32

1 **a** arc length = 8.7 cm
  area = 43.6 cm²
  b i $\frac{4}{3}\pi + 12$ cm **ii** 4π cm²
2 42 m³

Page 33

1 160π cm³ 2 112π cm²
3 605 cm³

Page 34

1 **i** 301.6 cm³ **ii** 188.5 cm²
2 **i** 36π cm³ **ii** 36π cm²
3 **i** 261.8 cm³ **ii** 235.6 cm²

Page 35

1 $3^2 + 4^2 = 5^2$
2 **a** 11.2 cm **b** 3.9 m

Page 36

1 36.7 cm² 2 7.07 cm

Page 37

1 **a i** 0.515 **ii** 0.574 **iii** 0.176
  b i 36.9° **ii** 45.6° **iii** 63.4°
2 30°; 8.34 cm; 7.52 cm

Page 38

1 **a** 5.3 cm **b** 32.2° **c** 11.3 cm
2 26.0 m

Page 39

1 **a** yes – a square has all the properties
    of a rectangle
  b yes – a rhombus has all the
    properties of a parallelogram
  c no – a kite does not have all side
    equal
2 **a i** 135° **ii** 45°
  b i 140° **ii** 40°
3 They always add up to 180°.

Page 40

1 **a** 80° – opposite angle in a cyclic
    quadrilateral
  b 160° – angle at centre twice angle at
    circumference
2 **a** 58° – sum of angles in a triangle
  b 32° – angle in same segment

Page 41

1 15° – angles in triangle are
  $5x + 5x + 2x = 180$
2 62° – angle in alternate segment is 56°
  and other two angles are equal
  (isosceles triangle)

Page 42

1 **a** A and C – SAS
  b B and D – D is 1.25 bigger than B
2 **a** $\binom{5}{0}$ **b** $\binom{-3}{4}$ **c** $\binom{5}{-4}$ **d** $\binom{0}{-4}$

Page 43

1 **a** y-axis **b** $x = 3$
2 **a** 90°, anticlockwise about (0, 0)
  b half-turn about (−1, 0)

Page 44

1 **a i** 2
  b i $\frac{2}{3}$
  b ii
  a ii

2 **a** reflection in y-axis
  b reflection is x-axis
  c 180° rotation about (0, 0)

Page 45

1 **a–c** self-checking
2 accurate drawing

Page 46

1–3 self-checking

Page 47

1 self-checking
2

•Edinburgh

•London

Page 48

1 20 m

Page 49

1 **a** 3 **b** 9 **c** 27
2 18.5 kg

Page 50

1 lengths: **a** and **e**, area: **c**, volumes: **b**, **d**
  and **f**
2 **a** A **b** N **c** N
  d V **e** N **f** L

Page 51

1 **a i** $a + b$ **ii** $3a$ **iii** $2a + 2b$
  iv $-2a + b$ **v** $-a - 2b$
  b $\overrightarrow{AT}$ is three times $\overrightarrow{BH}$ and parallel to
    it.
  c They are the same magnitude and in
    opposite directions.

Page 52

1 **a i** $-a + b$ or $b - a$
    ii $\frac{1}{3}a + \frac{1}{3}b$

  b $\overrightarrow{AP} = -a + 2b$ which is $2\overrightarrow{AM}$, hence
    $\overrightarrow{AP}$ and $\overrightarrow{AM}$ are multiples of each
    other and share a common point, so
    AMP is a straight line.

Page 55
1 **a** −7 **b** 25 **c** −28
2 **a** $3x + 15$ **b** $n^2 − 7n$ **c** $6p^3 − 9p^2q$
3 **a** $6x + 6$ **b** $10x + 38y$
4 **a** $3x(2x − 3)$
  b $2ab(a − 4 + 3b)$

Page 56
1 **a** $x = 3$ **b** $m = −1$
  c $n = 12$ **d** $x = \frac{1}{2}$
  e $x = 13$ **f** $x = 2$
  g $y = 13$ **h** $x = 28$ **i** $x = \frac{1}{2}$
2 **a** $x = 9$ **b** $m = 6$ **c** $x = −2$
  d $x = 4$ **e** $x = 2\frac{1}{2}$ **f** $y = −1\frac{1}{2}$
3 **a** $x = 7$ **b** $y = 2$ **c** $x = 3$
  d $x = −1$ **e** $x = \frac{1}{5}$ **f** $y = −2\frac{1}{2}$

Page 57
1 **a** $x = 9$ **b** $x = −3$ **c** $x = 5$
  d $x = 6\frac{1}{2}$
2 **a** 4 **b** 6
3 **a** $5x + 7 = 22$ **b** 3
4 **a** $\frac{x}{2} + 7 = x + 6$ **b** 2

Page 58
1 4.8
2 2.6

Page 59
1 **a** $x = 3, y = −2$
  b $x = 2\frac{1}{2}, y = 1\frac{1}{2}$
  c $x = 5, y = 1$
2 $x = 4, y = −1$

Page 60
1 **a** £7.50 **b** £5
2 **a** $x = \frac{T}{4}$ **b** $x = \frac{y − 3}{2}$
  c $x = P − 2t$ **d** $x = 5y$
  e $x = \frac{A − y}{m}$ **f** $x = \sqrt{\frac{S}{2\pi}}$

Page 61
1 **a** $x^2 + 2x − 3$
  b $m^2 − 4m − 12$
  c $n^2 − n − 2$
  d $x^2 + 6x + 5$
  e $x^2 − 6x + 9$
  f $x^2 + 10x + 21$
2 **a** $x^2 + 6x + 9$ **b** $x^2 − 4x + 4$
3 **a** $x^2 − 1$ **b** $m^2 − 4$

Page 62
1 **a** $(x + 3)(x − 1)$
  b $(x − 3)(x − 2)$
2 **a** $(x + 4)(x − 4)$
  b $(x + 6)(x − 6)$
3 **a** $(2x + 1)(x − 5)$
  b $(3x − 2)(2x + 3)$

4 **a** $x = −1$ or $x = 3$
  b $x = −5$ or $x = 2$

Page 63
1 **a** $x = −2$ or $x = \frac{3}{2}$
  b $x = −\frac{3}{4}$ or $x = −\frac{2}{3}$
2 **a** $x = ± \sqrt{\frac{7}{2}}$
  b $x = 0$ or $x = 3$
3 $x = −0.28$ or $x = 1.78$

Page 64
1 **a** $(x + 3)^2 − 10$
  b $(x − 4)^2 − 19$
2 **a** $−2 ± \sqrt{5}$ **b** $3 ± \sqrt{7}$
3 **a** because $b^2 − 4ac = −4 < 0$, it has no solutions
  b $−3 ± \sqrt{19}$

Page 65
1 **a** $3 ± \sqrt{\frac{13}{2}}$ **b** $−3 ± 2\sqrt{\frac{11}{2}}$
2 **a** $x − \frac{6}{x} = 1 \Rightarrow x^2 − 6 − x = 0$
  b $x = −2$ or $x = 3$
3 2 m

Page 66
1 **a** 20 minutes
  b car was stopped
  c B and C
  d 45 km/h
2 **a**

  b

Page 67
1 **a** 5.14 cm **b** 6.86 cm
  c 41.8°
2 35.3°

Page 68
1 **a** 18.9° **b** 54.2°

Page 69
1 **a** 23.6° and 156.4°
  b 84.3° and 275.7°
  c 224.4° and 315.6°
  d 113.6° and 246.4°

Page 70
1 **a** 69° **b** 12.6 cm
2 55.7° and 124.3°

Page 71
1 7.5 cm 2 112.4°

Page 72
1 **a** 10.6 cm **b** 16.0 cm
2 **a** 84.9° **b** 126.5°
3 24.2cm²

Page 73
1

2 4

Page 74
1 **a** A and C **b** A and B
2 **a**

  b

3 **a** **b**

Page 75
1 $y = 3x + 4$
2 **a** £6.50 **b** $1\frac{1}{2}$ **c** 2
  d $C = 1\frac{1}{2}d + 2$ **e** £6.50

Page 76
1 $x = 2, y = 3$
2 $y = −3x + 5$

Page 77

1 a (–3, 5), (–2, 1), (–1, –1),
(0, –1), (1, 1), (2, 5), (3, 11)

b

c –2.3, 1.3
d –1.6, 0.6

Page 78

1 a (0, –2)
b (–1.5, –4.25)
c 1, –3
d line is $y = -1$, solutions are 0.3, –3.3
e line is $y = x$, solutions are 0.7, –2.7

Page 79

1 Table: 0, 0.38, 0, 13.13, 24

2 Table: 5.66, 8, 11.3

Page 80

1 a $\frac{5x-1}{6}$ **b** $\frac{-7x-5}{15}$
c $\frac{3xy}{10}$ **d** $\frac{4}{3}$

Page 81

1 a $x = 5$
b $x = -\frac{3}{2}$ or $x = 3$
2 a $\frac{x+1}{2x+1}$ **b** $\frac{x-1}{2x-3}$

Page 82

1 a (4, 3), (–3, –4)
b (3, 2), (–2, 3)

Page 83

1 a i 4, 9, 14
ii 499
b i 3, 6, 10
ii 20 100
2 a $5n + 1$ **b** $8n - 5$ **c** $3n + 6$
3 a 200 **b** 199 **c** 10 000
d 28, 36, 45, 55
e 128, 256, 512, 1024
f 23, 29, 31, 37, 41

Page 84

1 a 41 **b** $4n + 1$
2 a $x = \frac{2y}{a-2}$ **b** $x = \frac{4+2y}{y-1}$

Page 85

1 a $x < 3$ **b** $x > 1$ **c** $x \geqslant 18$
d $x > -10$
2 a $x > 1$ **b** $x \leqslant 3$
3 a $x < 2$ **b** $x \geqslant -1$
c –1, 0, 1, 2, 3

Page 86

1 a **b** **c**

2

Page 87

1

2

3

Page 88

1

2

3

Page 89

1

Using triangle ABC,
angle ACB
$= 180 - (x + y)$
Angle ACB $= 180 - z$ (angles
on a straight line)
Hence $180 - (x + y) = 180 - z$
Hence $x + y = z$

2 $(n + 6)^2 - (n + 2)^2$
$= n^2 + 12n + 36 - (n^2 + 4n + 4)$
$= n^2 + 12n + 36 - n^2 - 4n - 4$
$= n^2 - n^2 + 12n - 4n + 36 - 4$
$= 8n + 32$
$= 8(n + 4)$

Index

Collins Revision

GCSE Higher
Maths

Exam Practice Workbook

FOR EDEXCEL A
FOR AQA A
FOR AQA B

Keith Gordon

Statistics 1

1 The table shows the number of cars per house on a housing estate of 100 houses.

Number of cars	Number of houses
0	8
1	23
2	52
3	15
4	2

a Find the modal number of cars.

_____ **[1 mark]**

b Find the median number of cars.

_____ **[1 mark]**

c Calculate the mean number of cars.

_____ **[2 marks]**

D

2 The table shows the results of a survey in which 50 boys were asked how many brothers and sisters they had.

		Number of brothers				
		0	1	2	3	4
Number of sisters	0	6	8	3	0	0
	1	3	5	7	1	0
	2	1	2	6	2	0
	3	1	0	1	2	1
	4	0	0	1	0	0

a What is the modal number of sisters? _____ **[1 mark]**

b What is the median number of sisters? _____ **[1 mark]**

c Calculate the mean number of brothers. _____ **[2 marks]**

d Mary says that the boys in this survey have more brothers than sisters. Explain why Mary is correct.

_____

_____ **[1 mark]**

C

This page tests you on   • averages • frequency tables

C

1 The table shows the scores of 200 boys in a mathematics examination.

The frequency polygon shows the scores of 200 girls in the same examination.

Mark, x	Frequency, f
$40 < x \leqslant 50$	27
$50 < x \leqslant 60$	39
$60 < x \leqslant 70$	78
$70 < x \leqslant 80$	31
$80 < x \leqslant 90$	13
$90 < x \leqslant 100$	12

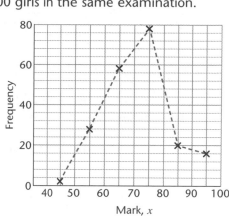

a Calculate the mean score for the boys.

_____ **[2 marks]**

b Draw the frequency polygon for the boys' scores on the same axes as the girls' scores. **[1 mark]**

c Who did better in the test, the boys or the girls? Give reasons for your answer.

_____

_____ **[1 mark]**

A

2 The table shows the lengths of 100 cucumbers.

Length, x (cm)	Frequency	Frequency density
$10 < x \leqslant 20$	15	
$20 < x \leqslant 25$	24	
$25 < x \leqslant 30$	36	
$30 < x \leqslant 40$	18	
$40 < x \leqslant 60$	7	

a Complete the table by calculating the frequency density for each class interval. **[1 mark]**

b Draw a histogram to show this information. **[2 marks]**

This page tests you on • grouped data and frequency diagrams • histograms

1 The table shows the units of gas used by one household over a three-year period and some of the four-point moving averages.

	2004				**2005**				**2006**			
Quarter	1st	2nd	3rd	4th	1st	2nd	3rd	4th	1st	2nd	3rd	4th
Units	230	182	140	200	210	174	136	192	198	158	120	180
4-point moving average			183	181	180	178	175	171	167	164		

a Explain why a four-point moving average is appropriate.

_____ **[1 mark]**

b Calculate the first four-point moving average.

[1 mark]

c Plot the four-point moving averages on these axes.

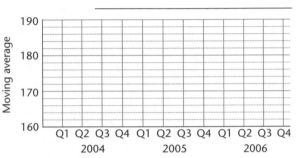

d Use the trend of the averages to predict the amount of gas used in the first quarter of 2007.

[1 mark]

_____ **[1 mark]**

2 Danny wanted to investigate the hypothesis: '*Girls spend more time on their mathematics coursework than boys do*'.

a Design a two-way table that Danny can use to record his data.

[2 marks]

b Danny collected data from 30 boys and 10 girls.

He found that, on average, the boys spent 10 hours on their mathematics coursework and the girls spent 11 hours on theirs. Does this prove the hypothesis?

Give reasons for your answer.

_____

_____ **[1 mark]**

This page tests you on • moving averages • surveys

C

1 This graph shows the price index
of petrol from 2000 to 2007.
In 2000 the price of a litre of
petrol was 60p.

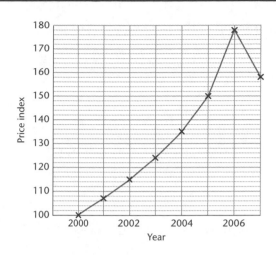

a Use the graph to find the price of a litre of petrol in 2007.

_____ **[1 mark]**

b One year there was an oil crisis as a result of a war in Iraq. What year do you
think it was?

Give a reason for your answer.

_____ **[1 mark]**

c Over the same seven-year period, the index for the cost of living rose by 38%.

Did the cost of petrol increase more quickly or more slowly than the cost of
living? Explain your answer.

_____

_____ **[1 mark]**

A

2 A national grocery chain wants to build a supermarket in a small market town.

One morning they asked 50 people who were shopping on the high street whether
they ever shopped in supermarkets. 32 of them said that they did.

At the planning meeting, the grocery chain claimed that 7 out of 10 of the people
in the town wanted the supermarket.

a Give two criticisms of the sampling method.

Criticism 1 _____ **[1 mark]**

Criticism 2 _____ **[1 mark]**

b Why is the conclusion invalid?

_____ **[1 mark]**

This page tests you on • social statistics • sampling

Statistics 2

1 This line graph shows the temperature in a greenhouse throughout one morning.

D

a Between which two hours was the biggest increase in temperature?

_____ **[1 mark]**

b The gardener opened the ventilator to lower the temperature. At what time did he do this?

_____ **[1 mark]**

c What was the approximate temperature at 10:30am?

_____ **[1 mark]**

d Can you use the graph to predict the temperature at 6pm? Explain your answer.

_____

_____ **[1 mark]**

---

2 A teacher recorded how many times the students in her form were late during a term.

C

The stem-and-leaf diagram shows the data.

12 students were **never late**.

0	2 3 4 4 5 6 7
1	3 5 8 9 9
2	0 1 4 5
3	2
5	1

Key: 1 | 7 represents 17

a How many students were in her form altogether?

_____ **[1 mark]**

b Work out the mean number of lates for the **whole form**.

_____ **[2 marks]**

This page tests you on • line graphs • stem-and-leaf diagrams

Scatter diagrams

1 A delivery driver records the distances and times for deliveries.

The table shows the results.

Delivery	Distance (km)	Time (minutes)
A	12	30
B	16	42
C	20	55
D	8	15
E	18	40
F	25	60
G	9	45
H	15	32
I	20	20
J	14	35

a Plot the values on the scatter diagram.

Distance (kilometres)

[2 marks]

b i During one of the deliveries the driver was stuck in a traffic jam.
   Which delivery was this?

_____ **[1 mark]**

ii One of the deliveries was done very early in the morning when there was
   no traffic. Which delivery was this?

_____ **[1 mark]**

c Ignoring the two values in **b**, draw a line of best fit through the rest of
   the data. **[1 mark]**

d Under normal conditions, how long would you expect a delivery of 22 kilometres
   to take?

_____ **[1 mark]**

This page tests you on • scatter diagrams • correlation • line of best fit

Cumulative frequency and box plots

1 The cumulative frequency diagram shows the amount of time 80 shoppers waited in a supermarket queue before being served.

a What is the median queuing time?

_____ **[1 mark]**

b What is the interquartile range?

_____ **[1 mark]**

c Use the graph to estimate the number of customers who spent less than $3\frac{1}{2}$ minutes in the queue.

_____ **[1 mark]**

B

2 The box plots show the marks students in two forms, 10K and 10J, achieved in the same geography test.

a What was the lowest mark in form 10K?

_____ **[1 mark]**

b What was the median mark for form 10J?

_____ **[1 mark]**

c Which form did better in the test? Give a reason for your answer.

_____

_____ **[1 mark]**

d Which form was more consistent in the test? Give a reason for your answer.

_____

_____ **[1 mark]**

B

This page tests you on • cumulative frequency diagrams • box plots

Probability

C

1 John made a dice and weighted one side with a piece of sticky gum.

He threw it 120 times. The table shows results.

Score	1	2	3	4	5	6
Frequency	18	7	22	21	35	17
Relative frequency						

a Complete the table by calculating the relative frequency of each score.

Give your answers to 2 decimal places.

[2 marks]

b On which side did John stick the gum?

Give a reason for your answer.

_____ **[1 mark]**

C

2 A bag contains 30 balls that are either red or white. The ratio of red balls to white balls is 2 : 3.

a Explain why the events 'picking a red ball at random' and 'picking a white ball at random' are mutually exclusive.

_____

_____ **[1 mark]**

b Explain why the events 'picking a red ball at random' and 'picking a white ball at random' are exhaustive.

_____

_____ **[1 mark]**

c Zoe says that the probability of picking a red ball at random from the bag is $\frac{2}{3}$. Explain why Zoe is wrong.

_____

_____ **[1 mark]**

d How many red balls are there in the bag?

_____ **[1 mark]**

e A ball is drawn at random from the bag, its colour is noted and then it is replaced in the bag.

This is repeated 200 times. How many of the balls would you expect to be red?

_____ **[1 mark]**

This page tests you on
- relative frequency
- mutually exclusive and exhaustive events

1 The two-way table shows the gender and departments of 40 teachers in a school.

	Male	Female
Mathematics	7	5
Science	11	7
RE	1	3
PE	3	3

a How many male teachers are there? _____ [1 mark]

b Which subject has equal numbers of male
and female teachers? _____ [1 mark]

c Nuna says that science is a more popular subject than mathematics for female
teachers.

Explain why Nuna is wrong.

_____

_____ [1 mark]

d What is the probability that a teacher chosen at random will be a female teacher
of mathematics or science?

_____ [1 mark]

D

2 The table shows the probability, by gender, of right- or left-handedness
of a student picked at random from a school.

	Boys	Girls
Right-handed	0.36	0.4
Left-handed	0.1	0.14

a What is the probability that a student picked at random from the school is
left-handed?

_____ [1 mark]

b There are 432 right-handed boys in the school.

How many students are there in the school altogether?

_____ [1 mark]

c In the town that the school serves, there are 26 000 people.

Estimate how many of them are left-handed.

_____ [1 mark]

C

This page tests you on • expectation • two-way tables
• addition rule for events

1 A fair 4-sided spinner is spun twice.

If the two scores are odd, the scores are added.

If one or both of the scores is even, the two scores are multiplied.

a Complete the table.

		First score			
		1	2	3	4
Second score	1	2	2	4	
	2				
	3				
	4				

[2 marks]

b What is the probability that the combined score will be an even number?

_____ [1 mark]

c What is the probability that the combined score will be a square number?

_____ [1 mark]

---

2 A bag contains three red balls and seven blue balls. A ball is taken out at random and replaced.

Another ball is then taken out.

a Complete the tree diagram to show the probabilities of the possible outcomes.

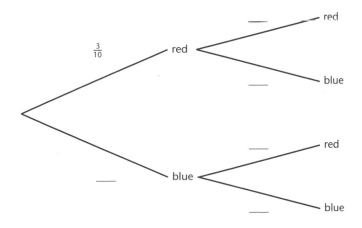

[2 marks]

b What is the probability that both balls are red?

_____ [1 mark]

c What is the probability that the balls are different colours?

_____ [2 marks]

This page tests you on • combined events • tree diagrams

1 A bag contained red and blue counters in the ratio 1 : 4.

a What is the probability of picking a red counter at random from the bag?

_____ **[1 mark]**

b There were six red counters in the bag. How many blue counters were there?

_____ **[1 mark]**

c Some red counters were added to the bag so that the probability of picking a red counter was $\frac{1}{2}$.

How many red counters were added?

_____ **[1 mark]**

B

2 The probability of throwing a 6 with a biased dice is x.

a What is the probability of **not** throwing a 6?

_____ **[1 mark]**

b The dice is thrown three times.

What is the probability of throwing three 6s?

_____ **[1 mark]**

B-A

3 A box contains two brown eggs and four white eggs. Mandy uses two eggs to make an omelette.

a Complete the tree diagram to show the probabilities of the possible outcomes.

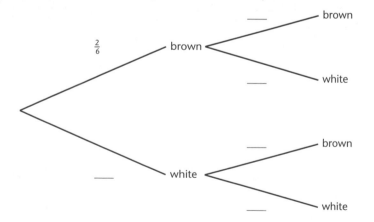

[2 marks]

b What is the probability she uses two eggs of the same colour?

_____ **[1 mark]**

A

This page tests you on • independent events • at least problems
• conditional probability

Handling data checklist

I can…

☐ draw an ordered stem-and-leaf diagram

☐ find the mean of a frequency table of discrete data

☐ draw a frequency polygon for discrete data

☐ find the mean from a stem-and-leaf diagram

☐ predict the expected number of outcomes of an event

☐ draw a line of best fit on a scatter diagram

☐ recognise the different types of correlation

☐ design a data collection sheet

☐ use the total probability of 1 to calculate the probabilities of events

You are working at ⬭ Grade D ⬭ level.

☐ find an estimate of the mean from a grouped table of continuous data

☐ draw a frequency diagram for continuous data

☐ calculate the relative frequency of an event from experimental data

☐ interpret a scatter diagram

☐ use a line of best fit to predict values

☐ design and criticise questions for questionnaires

You are working at ⬭ Grade C ⬭ level.

☐ calculate an n-point moving average

☐ use a moving average to predict future values

☐ draw a cumulative frequency diagram

☐ find the median, quartiles and interquartile range from a cumulative frequency diagram

☐ draw and interpret box plots

☐ draw a tree diagram to work out probabilities of combined events

You are working at ⬭ Grade B ⬭ level.

☐ draw histograms from frequency tables with unequal intervals

☐ calculate the numbers to be sampled for a stratified sample

☐ use AND/OR to work out the probabilities of combined events

You are working at ⬭ Grade A ⬭ level.

☐ find the median, quartiles and interquartile range from a histogram

☐ work out the probabilities of combined events when the probabilities change depending on previous outcomes (conditional probability)

You are working at ⬭ Grade A* ⬭ level.

Number

1 a The school canteen has 34 tables that each seat 14 students and 12 tables that each seat 8 students. What is the maximum number of students that can sit in the canteen at the same time?

D

_____ **[2 marks]**

b The head wants to expand the canteen so that it can cater for 700 students.

He wants to purchase an equal number of 12-seater and 8-seater tables.

How many extra tables will he need?

_____ **[2 marks]**

2 a Work these out.

i $10.8 \div 0.6$

_____ **[1 mark]**

ii $8.64 \div 0.36$

D

_____ **[1 mark]**

b A dividend is a repayment made every few months on the amount spent.

The Co-op pays a dividend of 2.6 pence for every £1 spent.

i In three months, Derek spends £240. How much dividend will he get?

_____ **[1 mark]**

ii Doreen received a dividend of £7.80 How much did she spend to get this dividend?

_____ **[1 mark]**

This page tests you on • real-life problems • dividing by decimals

C

1 a Round each number to the number of significant figures (sf) indicated.

    i 67 800 (1 sf)

    _____ **[1 mark]**

    ii 0.067 42 to (2 sf)

    _____ **[1 mark]**

b Find approximate answers to the following.

    i $\dfrac{312 \times 7.92}{0.42}$

    _____ **[1 mark]**

    ii $\dfrac{487}{0.52 \times 3.98}$

    _____ **[1 mark]**

D

2 a Write down the answers to:

    i 3.7×100 _____ **[1 mark]**

    ii 0.25×10^3 _____ **[1 mark]**

b Write down the answers to:

    i $7.6 \div 10$ _____ **[1 mark]**

    ii $0.65 \div 10^2$ _____ **[1 mark]**

c Write down the answers to:

    i $30\ 000 \times 400$ _____ **[1 mark]**

    ii 600^2 _____ **[1 mark]**

d Write down the answers to:

    i $90\ 000 \div 30$ _____ **[1 mark]**

    ii $30\ 000 \div 60$ _____ **[1 mark]**

This page tests you on
- rounding and approximating
- multiplying and dividing by powers of 10
- multiplying and dividing by multiples of 10

1 a You are told that p and q are prime numbers.

$p^2q^2 = 36$

What are the values of p and q?

$p =$ _____

$q =$ _____   **[2 marks]**

b Write 360 as the product of its prime factors.

_____   **[1 mark]**

c You are told that a and b are prime numbers.

$ab^2 = 98$

What are the values of a and b?

$a =$ _____

$b =$ _____   **[2 marks]**

d Write 196 as the product of its prime factors.

_____   **[1 mark]**

C

2 a Write 24 as the product of its prime factors.

_____   **[1 mark]**

b Write 60 as the product of its prime factors.

_____   **[1 mark]**

c What is the lowest common multiple of 24 and 60?

_____   **[1 mark]**

d What is the highest common factor of 24 and 60?

_____   **[1 mark]**

e In prime factor form, the number $P = 2^4 \times 3^2 \times 5$ and the number $Q = 2^2 \times 3 \times 5^2$.

i What is the lowest common multiple of P and Q?

Give your answer in index form.

_____   **[1 mark]**

ii What is the highest common factor of P and Q?

Give your answer in index form.

_____   **[1 mark]**

D-C

This page tests you on   • **prime factors**
• **lowest common multiple and highest common factor**

Fractions

C

1 a Work out $\frac{3}{4} + \frac{2}{5}$.

Give your answer as a mixed number.

_____ **[2 marks]**

b Work out $3\frac{2}{3} - 1\frac{4}{5}$.

Give your answer as a mixed number.

_____ **[2 marks]**

c On an aeroplane, two-fifths of the passengers are British, one-quarter are German, one-sixth are American and the rest are French.

What fraction of the passengers is French?

_____ **[2 marks]**

C

2 a Work out $2\frac{1}{2} \times 1\frac{2}{5}$.

Give your answer as a mixed number.

_____ **[2 marks]**

b Work out $3\frac{3}{10} \div 2\frac{2}{5}$.

Give your answer as a mixed number.

_____ **[2 marks]**

c Work out the area of this triangle.

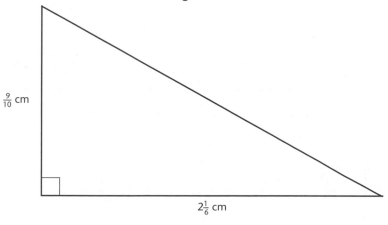

$\frac{9}{10}$ cm

$2\frac{1}{6}$ cm

_____ cm² **[2 marks]**

This page tests you on
• one quantity as a fraction of another
• adding and subtracting fractions
• multiplying and dividing fractions

Percentage

1 a A car cost £6000 new.

D

It depreciates in value by 12% in the first year and by 10% in the second year.

Which of these calculations shows the value of the car after 2 years?

Circle the correct answer.

6000 × 0.78   6000 × 0.88 × 0.9   6000 × 88 × 90   6000 − 2200   **[1 mark]**

b VAT is charged at 17.5 %.

A quick way to work out the VAT on an item is to work out 10%, then divide this by 2 to get 5%, then divide this by 2 to get 2.5%. Finally, add these three values together.

Use this method to work out the VAT on an item costing £68.

_____ **[2 marks]**

2 a A computer costs £700 excluding VAT.

D

VAT is charged at 17.5%.

What is the cost of the computer when VAT is added?

_____ **[2 marks]**

b The price of a printer is reduced by 12% in a sale.

The original price of the printer was £250.

What is the price of the printer in the sale?

_____ **[2 marks]**

3 In its first week, a new bus route carried a total of 2250 people.

C

In its second week it carried 2655 people.

What is the percentage increase in the number of passengers from the first week to the second?

_____ **[2 marks]**

This page tests you on
- the percentage multiplier
- calculating a percentage increase or decrease
- expressing one quantity as a percentage of another

C

1 a A car cost £8000 new.

It depreciates in value by 12% each year.

Which of these calculations shows the value of the car after 2 years?

Circle the correct answer.

8000 × 0.76   8000 × 0.88 × 0.88   8000 × 88 × 88   8000 − 2400   **[1 mark]**

b Mary invested £2000 in an account which paid 3.5% interest each year.

She left her money in the bank for six years.

How much money did she have at the end of this period?

_____ **[2 marks]**

c Over the same period, Bert invests £2000 in shares which lost 2% in value each year. How much were his shares worth after six years?

_____ **[2 marks]**

B

2 a Electro Co had a dishwasher priced at £240. Its price was reduced by 5% in a sale.

Corries had the same dishwasher priced at £260. Its price was reduced by 12% in a sale.

At which company was the dishwasher cheaper?

Show all your working clearly.

_____ **[2 marks]**

b In the same sale, Electro Co reduced the price of a cooker by 5% to £361.

What was the original price of the cooker?

_____ **[2 marks]**

This page tests you on   • compound interest • reverse percentage

Ratio

1 a Write the ratio 12 : 9 in its simplest form.

_____ **[1 mark]**

b Write the ratio 5 : 2 in the form 1 : n.

_____ **[1 mark]**

c A fruit drink is made from orange juice and cranberry juice in the ratio 5 : 3.

If 1 litre of the drink is made, how much of the drink is cranberry juice?

_____ **[2 marks]**

C

---

2 In a tutor group the ratio of girls to boys is 3 : 4. There are 15 girls in the form.

How many students are there in the form altogether?

_____ **[2 marks]**

C

---

3 A ferry covers the 72 kilometres between Holyhead and Dublin in $2\frac{1}{4}$ hours.

a What is the average speed of the ferry?

State the units of your answer.

_____ **[2 marks]**

b For the first 15 minutes and the last 15 minutes of the journey, the ferry is manoeuvring in and out of the ports. During this time the average speed is 18 km per hour.

What is the average speed of the ferry during the rest of the journey?

State the units of your answer.

_____ **[3 marks]**

C

This page tests you on • ratios • speed, time and distance

Grades

1 A car uses 50 litres of petrol in driving 275 miles.

   a How much petrol will the car use in driving 165 miles?

_____ **[2 marks]**

   b How many miles can the car drive on 26 litres of petrol?

_____ **[1 mark]**

D

2 Nutty Flake cereal is sold in two sizes.

   The handy size contains 600 g and costs £1.55.

   The large size contains 800 g and costs £2.20.

   Which size is the better value?

_____ **[2 marks]**

C

3 A block of metal has a volume of 750 cm^3 and a mass of 5.1 kg.

   Calculate the density of the metal.

   State the units of your answer.

_____ **[2 marks]**

This page tests you on • direct proportion problems • best buys • density

Powers and reciprocals

1 a Write down the value of:

   i $\sqrt{169}$

   _____ **[1 mark]**

   ii 5^3

   _____ **[1 mark]**

b Write down the value of:

   i $\sqrt[3]{64}$

   _____ **[1 mark]**

   ii 2^8

   _____ **[1 mark]**

D

2 a Fill in the missing numbers.

			last digit
4^1	=	4	4
4^2	=	16	6
4^3	=	___	___
4^4	=	___	___
4^5	=	___	___

[2 marks]

b What is the last digit of 4^{99}?

Explain your answer.

_____ **[1 mark]**

c Which is greater: 5^6 or 6^5?

Justify your answer.

_____ **[1 mark]**

D

3 Write each of these as a fraction in its simplest form.

  a 8^{-1}

   _____ **[1 mark]**

  b 4^{-2}

   _____ **[1 mark]**

  c $2^{-3} + 4^{-1}$

   _____ **[2 marks]**

B

This page tests you on   • square roots and cube roots   • powers
                                      • negative indices

B

1 a Write the number 45.2×10^3 in standard form.

_____ **[1 mark]**

b Write the number 0.6×10^{-2} as an ordinary number.

_____ **[1 mark]**

c Here are six numbers.

2.5×10^5 1.8×10^6 45.2×10^3 8×10^{-4} 4 4.8×10^{-1} 0.6×10^{-2}

Which of the six numbers is the largest?

_____ **[1 mark]**

d Which of the six numbers is the smallest?

_____ **[1 mark]**

e Work out $2.5 \times 10^5 \times 8 \times 10^{-4}$.

Give your answer in standard form.

_____ **[1 mark]**

f Work out $(4.8 \times 10^{-1}) \div (8 \times 10^{-4})$.

Give your answer in standard form.

_____ **[1 mark]**

C

2 a $\frac{1}{9} = 0.1111...$ $\frac{2}{9} = 0.2222...$

Use this information to write down the decimal equivalent of:

i $\frac{4}{9}$ _____ **[1 mark]**

ii $\frac{5}{9}$ _____ **[1 mark]**

b Write down the reciprocal of each number.

Give your answers as fractions or mixed numbers as appropriate.

i 10 _____ **[1 mark]**

ii $\frac{3}{4}$ _____ **[1 mark]**

c Write your answers to **b** as terminating or recurring decimals, as appropriate.

i _____ **[1 mark]**

ii _____ **[1 mark]**

d Work out the reciprocal of each number.

i 1.25 _____ **[1 mark]**

ii 2.5 _____ **[1 mark]**

iii 5 _____ **[1 mark]**

This page tests you on   • standard form • rational numbers
• finding reciprocals

1 a Write $x^5 \times x^2$ as a single power of x.

_____ [1 mark]

b Write $x^8 \div x^4$ as a single power of x.

_____ [1 mark]

c i If $3^n = 81$, what is the value of n?

_____ [1 mark]

ii If $3^m = 27$, what is the value of m?

_____ [1 mark]

d Write down the value of $2^2 \times 5^2 \times 2^4 \times 5^4$.

_____ [1 mark]

C

2 Simplify the following expressions.

a $\dfrac{8a^3b \times 4a^2b^4}{2ab^2}$

_____ [2 marks]

b $(3x^2y^3)^2$

_____ [2 marks]

B

3 Write down the value of each of these.

a $8^{\frac{1}{3}}$

_____ [1 mark]

b $121^{-\frac{1}{2}}$

_____ [1 mark]

c $81^{-\frac{3}{4}}$

_____ [2 marks]

A–A*

This page tests you on • multiplying and dividing powers • power of a power • indices of the form $\frac{1}{n}$ and $\frac{a}{b}$

Surds

A*

1 a Simplify $\sqrt{48}$ as far as possible by writing it in the form $a\sqrt{b}$, where a and b are integers.

_____ **[1 mark]**

b Write $\sqrt{48} + \sqrt{75}$ in the form $p\sqrt{q}$, where p and q are integers.

_____ **[1 mark]**

c Expand and simplify $(\sqrt{3} + 5)(\sqrt{3} - 1)$.

_____ **[2 marks]**

d Expand and simplify $(\sqrt{5} + 2)(\sqrt{5} - 2)$.

_____ **[2 marks]**

A*

2 a Rationalise the denominator of $\dfrac{3}{\sqrt{6}}$.

Simplify your answer as much as possible.

_____ **[2 marks]**

b This rectangle has an area of 20 cm².

Find the value of x.

Give your answer in surd form.

x cm

$5\sqrt{2}$

_____ cm **[1 mark]**

c Show clearly that $= \dfrac{4}{\sqrt{12}} + \dfrac{\sqrt{12}}{4} = \dfrac{7\sqrt{3}}{6}$.

_____ **[2 marks]**

This page tests you on • surds • rationalising the denominator
• solving problems with surds

123

Variation

1 The mass of a cube is directly proportional to the cube of its side.

Let m represent the mass of the cube.

Let s represent the side of the cube.

a Write down the proportionality equation that connects s and m.

_____ **[1 mark]**

b When $s = 10$, $m = 50$. Find the value of the constant of proportionality, k.

_____ **[1 mark]**

c Find the value of m when $s = 20$.

_____ **[1 mark]**

d Find the value of s when $m = 3.2$.

_____ **[1 mark]**

2 a Two variables, a and b, are known to be proportional to each other.

When $a = 2$, $b = 4$.

Find the constant of proportionality, k, if:

i $a \propto b^2$

_____ **[1 mark]**

ii $a \propto \dfrac{1}{\sqrt{b}}$

_____ **[1 mark]**

b y is inversely proportional to the cube root of x.

When $y = 10$, $x = 8$.

Find the value of y when $x = 125$.

_____ **[2 marks]**

This page tests you on • direct variation • inverse variation

Limits

C

1 a A coach is carrying 50 people, rounded to the nearest 10.

i What is the smallest number of people that could have been on the coach?

_____ **[1 mark]**

ii What is the greatest number of people that could have been on the coach?

_____ **[1 mark]**

b The coach is travelling at 50 mph, to the nearest 10 mph.

i What is the lowest speed the coach could have been doing?

_____ **[1 mark]**

ii What is the greatest speed the coach could have been doing?

_____ **[1 mark]**

A

2 A cube has a sides of 20 cm, measured to the nearest centimetre.

a What is the lowest possible value of the volume of the cube?

_____ **[2 marks]**

b What is the greatest possible value of the volume of the cube?

_____ **[2 marks]**

A*

3 x and y are continuous values, both measured to 2 significant figures.

$x = 230$ and $y = 400$.

Work out the greatest possible value of $\frac{x}{y^2}$.

_____ **[2 marks]**

This page tests you on • limits of accuracy
• calculating with limits of accuracy

Number checklist

I can...

- [ ] work out one quantity as a fraction of another
- [ ] solve problems using negative numbers
- [ ] multiply and divide by powers of 10
- [ ] multiply together numbers that are multiples of powers of 10
- [ ] round numbers to one significant figure
- [ ] estimate the answer to a calculation
- [ ] order lists of numbers containing decimals, fractions and percentages
- [ ] multiply and divide fractions
- [ ] calculate with speed, distance and time
- [ ] compare prices to find 'best buys'
- [ ] find the new value after a percentage increase or decrease
- [ ] find one quantity as a percentage of another

You are working at (Grade D) level.

- [ ] work out a reciprocal
- [ ] recognise and work out terminating and recurring decimals
- [ ] write a number as a product of prime factors
- [ ] find the HCF and LCM of pairs of numbers
- [ ] use the index laws to simplify calculations and expressions
- [ ] multiply and divide with negative and mixed numbers
- [ ] round numbers to given numbers of significant figures
- [ ] find a percentage increase
- [ ] work out compound interest problems
- [ ] solve problems using ratio

You are working at (Grade C) level.

- [ ] work out the square roots of decimal numbers
- [ ] estimate answers using square roots of decimal numbers
- [ ] work out reverse percentage problems
- [ ] solve problems involving density
- [ ] write and calculate with numbers in standard form
- [ ] find limits of numbers given to the nearest unit

You are working at (Grade B) level.

- [ ] solve complex problems involving percentage increases and decreases
- [ ] use the rules of indices for fractional and negative indices
- [ ] convert recurring decimals to fractions
- [ ] simplify surds
- [ ] find limits of numbers given to various accuracies

You are working at (Grade A) level.

- [ ] solve problems using surds
- [ ] solve problems using combinations of numbers rounded to various limits
- [ ] You are working at (Grade A*) level.

Circles and area

D

1 a Work out the area of a circle of radius 12 cm.

Give your answer to 1 decimal place.

_____ cm² **[2 marks]**

b Work out the circumference of a circle of diameter 20 cm.

Give your answer to 1 decimal place.

_____ cm **[2 marks]**

C

2 Work out the area of a semicircle of diameter 20 cm.

Give your answer in terms of π.

20 cm

_____ cm² **[2 marks]**

D

3 Calculate the area of this trapezium.

16 cm

8 cm

22 cm

_____ cm² **[2 marks]**

Sectors and prisms

A

1 OAB is a minor sector of a circle of radius 10 cm.

Angle AOB $= 72°$.

a Calculate the area of the **minor** sector OAB.

Give your answer in terms of π.

_____ cm^2

[2 marks]

b Calculate the perimeter of the **major** sector OAB.

Give your answer in terms of π.

_____ cm   **[2 marks]**

D

2 A triangular prism has dimensions as shown.

a Calculate the cross-sectional area of the prism.

_____ cm^2 **[1 mark]**

b Calculate the volume of the prism.

_____ cm^3 **[2 marks]**

This page tests you on   • sectors • prism

Cylinders and pyramids

C–A

1 A cylinder has a radius 4 cm and a height of 10 cm.

    a What is the volume of the cylinder?

       Give your answer in terms of π.

       _____ cm³   **[2 marks]**

    b What is the **total** surface area of the cylinder?

       Give your answer in terms of π.

       _____ cm²   **[2 marks]**

A*

2 The pyramid in the diagram has its top 3 cm cut off, as shown.

The shape which is left is called a **frustum**.

Calculate the volume of the frustum.

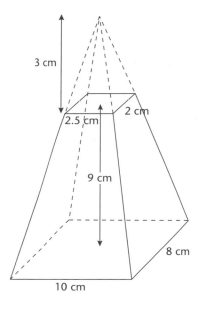

3 cm

2 cm

2.5 cm

9 cm

8 cm

10 cm

       _____ cm³   **[2 marks]**

This page tests you on   • **volume of a cylinder** • **surface area of a cylinder**
                              • **volume of a pyramid**

Cones and spheres

1 A fishing float is made of light wood with a density of 1.5 g/cm³.

The float is shaped as two cones, each of diameter 1 cm, joined at their bases.

One cone is 6 cm long and the other is 3 cm long.

Find the total mass of the float.

A

1 cm

6 cm

3 cm

_____ g   **[4 marks]**

2 A marble ashtray is made from a cylinder from which a hemisphere has been cut.

The cylinder has a radius of 6 cm and a depth of 6 cm.

The hemisphere has a radius of 5 cm.

What is the volume of the ashtray?

A

5 cm

6 cm

6 cm

_____ cm³   **[3 marks]**

This page tests you on   • cones • spheres

Pythagoras' theorem

C

1 Calculate the length of the side marked x in this right-angled triangle.

Give your answer to 1 decimal place.

_____ cm   **[2 marks]**

C

2 Calculate the length of the side marked x in this right-angled triangle.

Give your answer to 1 decimal place.

_____ cm   **[2 marks]**

C

3 A flagpole 4 m tall is supported by a wire that is fixed at a point 2.1 m from the base of the pole.

How long is the wire? (The length is marked x on the diagram.)

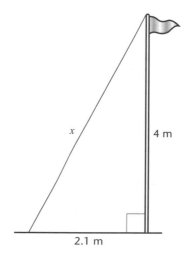

_____ m   **[2 marks]**

This page tests you on   • Pythagoras' theorem • finding lengths of sides
• real-life problems

1 Calculate the area of an isosceles triangle with sides of 12 cm, 12 cm and 7 cm.

12 cm   12 cm

7 cm

_____ cm² **[2 marks]**

2 The diagram shows a square-based pyramid with base length 10 cm and sloping edges 15 cm.

M is the mid-point of side AB, X is the mid-point of the base and E is directly above X.

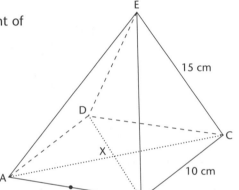

a Calculate the length of the diagonal AC.

_____ cm **[2 marks]**

b Calculate EX, the height of the pyramid.

_____ cm **[2 marks]**

c Using triangle ABE, calculate the length EM.

_____ cm **[2 marks]**

This page tests you on
- Pythagoras' theorem and isosceles triangles
- Pythagoras' theorem in three dimensions

Using Pythagoras and trigonometry

C

1 Use your calculator to work out sin (tan⁻¹ 0.65).

   a Write down all of the numbers in the calculator display.

   _____ **[1 mark]**

   b Round your answer to an appropriate degree of accuracy.

   _____ **[1 mark]**

B

2 Sine is defined as $\sin a = \dfrac{\text{opposite}}{\text{hypotenuse}}$.

   a Find the length of the side marked x in this right-angled triangle.

13 cm
x
40°

   _____ cm **[2 marks]**

   b Find the size of the angle marked y in this right-angled triangle.

15 cm
8 cm
y

   _____ ° **[2 marks]**

B

3 Cosine is defined as $\cos a = \dfrac{\text{adjacent}}{\text{hypotenuse}}$.

   a Find the length of the side marked x in this right-angled triangle.

x
48°
15 cm

   _____ cm **[2 marks]**

   b Find the size of the angle marked y in this right-angled triangle.

20 cm
y
9 cm

   _____ ° **[2 marks]**

B

4 Tangent is defined as $\tan a = \dfrac{\text{opposite}}{\text{adjacent}}$.

   a Find the length of the side marked y in this right-angled triangle.

20°
y
10 cm

   _____ cm **[2 marks]**

   b Find the size of the angle marked x in this right-angled triangle.

3 cm
x
5 cm

   _____ ° **[2 marks]**

This page tests you on • trigonometry • sine • cosine • tangent

1 Find the size of the angle marked x in this right-angled triangle.

B

_____ ° **[2 marks]**

2 Find the length of the side marked x in this right-angled triangle.

B

_____ cm **[2 marks]**

3 Find the length of the side marked x in this right-angled triangle.

B

_____ cm **[2 marks]**

4 This diagram shows a rack railway. It ascends to a height of 40 m using 50 m of track.

What angle does the track make with the horizontal?

B

_____ ° **[2 marks]**

This page tests you on • which ratio to use • solving problems

Geometry

C

1 The diagram shows a kite, ABCD, joined to a parallelogram, CDEF.

Angle BAD = 100° and angle CFE = 60°.

When the side of the kite is extended, it passes along the diagonal of the parallelogram shown by the dotted line.

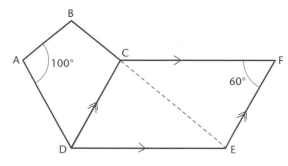

Use the properties of quadrilaterals to find the size of angle CED. Show all your working.

_____° **[4 marks]**

C

2 a Explain why the interior angles of a pentagon add up to 540°.

_____

_____

_____

_____

_____ **[2 marks]**

b The diagram shows three sides of a regular polygon.

The exterior angle is 36°.

How many sides does the polygon have altogether?

_____ **[2 marks]**

c The interior angle of a regular polygon is 160°.

Explain why the polygon must have 18 sides.

_____ **[2 marks]**

Circle theorems

1 In the diagram, O is the centre of the circle. ABC is a triangle, AB is a diameter of the circle and angle CBA = 63°.

B

a Find the size of angle CAB.

_____ °

Give a reason for your answer.

_____

_____ **[2 marks]**

b Explain why the angle OCB is 63°.

_____

_____ **[1 mark]**

2 In the diagram, O is the centre of the circle.

B

a State the value of x.

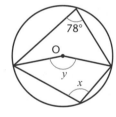

_____ °

Give a reason for your answer.

_____

_____ **[2 marks]**

b State the value of y.

_____ °

Give a reason for your answer.

_____

_____ **[2 marks]**

This page tests you on • circle theorems

A

1 In the diagram, PT is a tangent to the circle, centre O, at Q.

AB is a diameter of the circle and angle QBA = 57°.

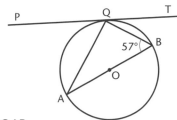

a Calculate the size of angle QAB.

∠QAB = _____ °

Give a reason for your answer.

_____

_____ **[2 marks]**

b Calculate the size of angle BQT.

Give reasons for your answers.

∠BQT = _____ °

_____

_____ **[2 marks]**

A*

2 The diagram shows a circle with centre O. ABC is an isosceles triangle with AB = AC.

PT is a tangent to the circle at A, M is the mid-point of BC and angle ABC = $x°$.

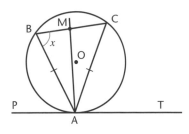

Explain why AM must pass through the centre of the circle.

Give reasons for any values or angles you write down.

_____

_____

_____

_____ **[2 marks]**

This page tests you on • tangents and chords • alternate segment theorem

Transformation geometry

1 Triangles A, P, Q, R, S and T are not drawn accurately.

C

Which two of triangles P, Q, R, S and T are congruent to triangle A?

_____ and _____ **[1 mark]**

2 a Describe the transformation that takes the shaded triangle to triangle A.

D-C

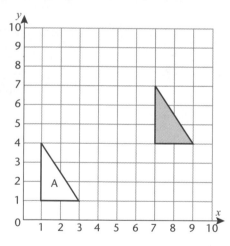

_____ **[2 marks]**

b Translate the shaded triangle by $\binom{-4}{2}$.

Label the image B. **[1 mark]**

c The shaded triangle is translated by $\binom{7}{-5}$ to give triangle C.

What vector will translate triangle C to the shaded triangle?

_____ **[1 mark]**

1

D

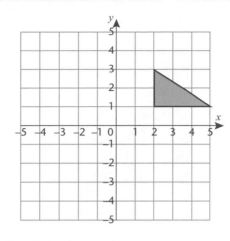

a Reflect the shaded triangle in the *y*-axis.
Label the image A.   [1 mark]

b Reflect the shaded triangle in the line $y = -1$.
Label the image B.   [1 mark]

c Reflect the shaded triangle in the line $y = -x$.
Label the image C.   [1 mark]

2

D

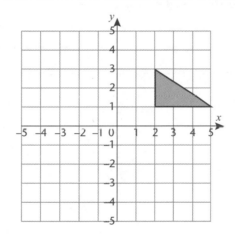

a Rotate the shaded triangle by 90° clockwise about (1, 0).
Label the image A.   [1 mark]

b Rotate the shaded triangle by a half-turn about (1, 3).
Label the image B.   [1 mark]

c What rotation will take triangle A to triangle B?

_____   [3 marks]

This page tests you on   • reflections • rotations

1

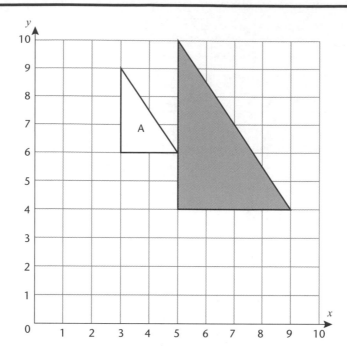

C

a What transformation takes the shaded triangle to triangle A?

_____ **[2 marks]**

b Draw the image after the shaded triangle is enlarged
by a scale factor $\frac{1}{4}$ about (1, 0). **[1 mark]**

2

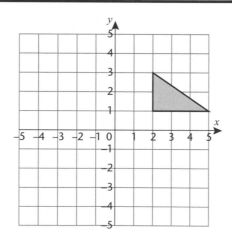

C

a Rotate the shaded triangle 90° clockwise about the origin.
Label the image A. **[1 mark]**

b Reflect triangle A in the y-axis.
Label the image B. **[1 mark]**

c What **single** transformation maps the shaded triangle to triangle B?

_____ **[1 mark]**

This page tests you on • enlargements • combined transformations

Constructions

Grades

D

1 Make an accurate drawing of this triangle.

[3 marks]

C

2 a Using compasses and a ruler, construct an angle of 60° at the point A.

b i Use compasses and a ruler to construct this triangle accurately.

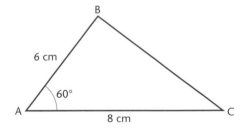

[3 marks]

ii Measure the length of the line BC. _____ cm **[1 mark]**

This page tests you on • constructing triangles • constructing an angle of 60°

SHAPE, SPACE AND MEASURES

1 Use compasses and a ruler to do these constructions.

   a Construct the perpendicular bisector of AB.

A ●

● B

[2 marks]

   b Construct the perpendicular
      at the point C to the line L.

────────────────●──────────────── L
                       C

[2 marks]

C

2 Use compasses and ruler to do these constructions.

   a Construct the perpendicular bisector of the line L.

──────────────────── L

[2 marks]

   b Construct the bisector of angle ABC.

[2 marks]

C

This page tests you on   • the perpendicular bisector   • the angle bisector
                                   • the perpendicular at a point on a line

Constructions and loci

C

1 Use compasses and ruler to construct
the perpendicular from the point C
to the line L.

• C

——————————————————————— L

[2 marks]

C

2 The diagram, which is drawn to scale, shows a flat, rectangular lawn of length
10 m and width 6 m, with a circular flower bed of radius 2 m.

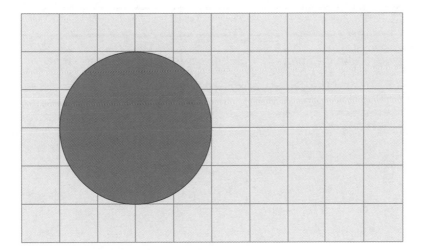

A tree is going to be planted in the garden.

It has to be at least 1 metre from the edge of the garden and at least 2 metres
from the flower bed.

a Draw a circle to show the area around the flower bed where the tree *cannot*
be planted.   [1 mark]

b Show the area of the garden in which the tree *can* be planted.   [1 mark]

This page tests you on   • the perpendicular from a point to a line   • loci
• practical problems

Similarity

B

1 Triangle ABC is similar to triangle CDE.

a Find the length of CD.

_____ cm   **[1 mark]**

b Find the length of AC.

_____ cm   **[1 mark]**

A

2 A pinhole camera projects an image through a pinhole onto light-sensitive paper.

The paper is 12 cm by 8 cm and the camera box is 20 cm deep.

When the subject of the photo is 15 metres from the front of the camera, what are the width and depth of the total area that will be photographed?

width _____ m

depth _____ m   **[2 marks]**

This page tests you on   • similar triangles   • special cases of similar triangles

A

1 In a 'home' snooker set, everything is made to three-quarters of the size of regular snooker equipment.

   a A full-size snooker table is 3.6 m by 1.8 m.

      What is the size of a 'home' snooker table?

      _____ m by _____ m   **[2 marks]**

   b The manufacturers recommend a room with an area of 52 m² for a regular snooker table.

      What room area should be recommended for a 'home' snooker set?

      _____ m²   **[2 marks]**

   c Regular snooker balls have a volume of 75 cm³. What is the volume of a 'home' snooker ball?

      _____ cm³   **[2 marks]**

A*

2 Ecowash washing powder is sold in two sizes: standard and family.

It is sold in cuboidal boxes that are similar in shape.

The family size contains 2.5 kg of powder.

The standard size contains 1 kg of powder.

The amount of powder is proportional to the volume of the boxes.

   a The height of the standard-size box is 21 cm.

      What is the height of the family-size box?

      _____   **[2 marks]**

   b The labels on the front of each box are also similar in shape.

      The area of the label on the family-size box is 70 cm².

      What is the area of the label on the standard-size box?

      _____   **[2 marks]**

This page tests you on
- similar shapes
- solving problems with area and volume ratios

Dimensional analysis

1 A cuboid has sides of length x, y and z cm.

One of the following formulae represents the total lengths of the edges (L).

One of them represents the total area of the faces (A).

One of them represents the total volume (V).

Indicate which is which.

$2xz + 2yz + 2xy$ represents total _____ **[1 mark]**

xyz represents total _____ **[1 mark]**

$4x + 4y + 4z$ represents total _____ **[1 mark]**

---

2 A can of beans is a cylinder with a radius r cm and a height h cm.

The can has a label around it that is glued together with an overlap of 1 centimetre.

a One of these formulae represents the perimeter of the label (P).

One of them represents the area of the two ends of the can (A).

One of them represents the volume of the can (V).

Indicate which is which.

$2\pi r^2$ represents the _____ **[1 mark]**

$4\pi r + 2h + 2$ represents the _____ **[1 mark]**

$\pi r^2 h$ represents the _____ **[1 mark]**

b Adil works out a formula for the total surface area of the can as $2\pi r^2 + 2\pi r^2 h$.

Explain why this cannot be a correct formula.

_____

_____ **[1 mark]**

This page tests you on   • dimensional analysis

Vectors

1 On this grid $\overrightarrow{OA}$ = **a** and $\overrightarrow{OB}$ = **b**.

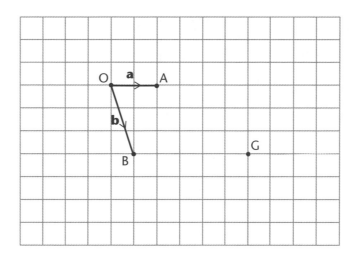

a Label each of the following points on the grid.

   i C such that:
     $\overrightarrow{OC}$ = 2**b**

   ii D such that:
      $\overrightarrow{OD}$ = 2**a** + **b**

   iii E such that:
       $\overrightarrow{OE}$ = –**a**

   iv F such that:
      $\overrightarrow{OF}$ = 2**b** – 2**a**   **[4 marks]**

b Give OG in terms of **a** and **b**.

                                                            [1 mark]

c The point H, which is not on the diagram, is such that
  $\overrightarrow{OH}$ = –5**a** + 7**b**.

   Write down the vector $\overrightarrow{HO}$.

                                                              [1 mark]

This page tests you on   • vectors • addition and subtraction of vectors

1 OABC is a quadrilateral.

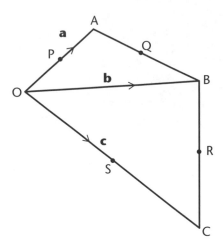

$\overrightarrow{OA}$ = **a**, $\overrightarrow{OB}$ = **b** and $\overrightarrow{OC}$ = **c**.

The mid-points of OA, AB, BC and OC are P, Q, R and S respectively.

a Give the following vectors in terms of **a**, **b** and **c**.

i $\overrightarrow{PS}$ _____ [1 mark]

ii $\overrightarrow{AB}$ _____ [1 mark]

iii $\overrightarrow{BC}$ _____ [1 mark]

iv $\overrightarrow{QR}$ _____ [1 mark]

b What type of quadrilateral is PQRS?

Explain your answer.

_____

_____

_____

_____ [1 mark]

A*

Shape, space and measures checklist

I can...

☐ find the area of a parallelogram using the formula $A = bh$

☐ find the area of a trapezium using the formula $\frac{1}{2}(a + b)h$

☐ find the area of a compound shape

☐ work out the formula for the perimeter, area or volume of simple shapes

☐ identify the planes of symmetry for 3-D shapes

☐ recognise and find alternate angles in parallel lines and a transversal

☐ recognise and find corresponding angles in parallel lines and a transversal

☐ recognise and find interior angles in parallel lines and a transversal

☐ use and recognise the properties of quadrilaterals

☐ find the exterior and interior angles of regular polygons

☐ understand the words 'sector' and 'segment' when used with circles

☐ calculate the circumference of a circle, giving the answer in terms of π if necessary

☐ calculate the area of a circle, giving the answer in terms of π if necessary

☐ recognise plan and elevation from isometric and other 3-D drawings

☐ translate a 2-D shape

☐ reflect a 2-D shape in lines of the form $y = a$ and $x = b$

☐ rotate a 2-D shape about the origin

☐ enlarge a 2-D shape by a whole number scale factor about the origin

☐ construct diagrams accurately using compasses, a protractor and a straight edge

☐ use appropriate conversion factors to change between imperial and metric units and vice versa

You are working at ⬭ Grade D ⬭ level.

☐ work out the formula for the perimeter, area or volume of complex shapes

☐ work out whether an expression or formula represents a length, an area or a volume

☐ relate the exterior and interior angles in regular polygons to the number of sides

☐ find the area and perimeter of semicircles

☐ translate a 2-D shape, using a vector

☐ reflect a 2-D shape in the lines $y = x$ and $y = -x$

☐ rotate a 2-D shape about any point

☐ enlarge a 2-D shape by a fractional scale factor

☐ enlarge a 2-D shape about any centre

☐ construct perpendicular and angle bisectors

☐ construct an angle of 60°

☐ construct the perpendicular to a line from a point on the line and from a point to a line

☐ draw simple loci

☐ work out the surface area and volume of a prism

☐ work out the volume of a cylinder using the formula $V = \pi r^2 h$

☐ find the density of a 3-D shape

☐ find the hypotenuse of a right-angled triangle, using Pythagoras' theorem

- [ ] find the short side of a right-angled triangle, using Pythagoras' theorem
- [ ] use Pythagoras' theorem to solve real-life problems

You are working at (Grade C) level.

- [ ] calculate the length of an arc and the area of a sector
- [ ] use trigonometric ratios to find angles and sides in right-angled triangles
- [ ] use trigonometry to solve real life problems involving right-angled triangles
- [ ] use circle theorems to find angles, in circle problems
- [ ] use the conditions for congruency to identify congruent triangles
- [ ] enlarge a shape by a negative scale factor
- [ ] use similar triangles to find missing lengths
- [ ] identify whether a formula is dimensionally consistent, using dimensional analysis
- [ ] understand simple proofs such as the exterior angle of a triangle is equal to the sum of the opposite interior angles

You are working at (Grade B) level.

- [ ] calculate the surface area of cylinders, cones and spheres
- [ ] calculate the volume of cones and spheres
- [ ] solve 3-D problems using Pythagoras' theorem
- [ ] use the alternate segment theorem to find angles in circle problems
- [ ] prove that two triangles are congruent
- [ ] solve more complex loci problems
- [ ] solve problems using area and volume scale factors
- [ ] solve real-life problems, using similar triangles
- [ ] solve problems using addition and subtraction of vectors
- [ ] use the sine rule to solve non right-angled triangles
- [ ] use the cosine rule to solve non right-angled triangles
- [ ] use the rule $A = \frac{1}{2}ab\sin C$ to find the area of non right-angled triangles

You are working at (Grade A) level.

- [ ] calculate the volume and surface area of compound 3-D shapes
- [ ] use circle theorems to prove geometrical results
- [ ] solve more complex problems using the proportionality of area and volume scale factors
- [ ] solve problems using vector geometry
- [ ] use the sine rule to solve more complex problems, involving right-angled and non right-angled triangles
- [ ] use the cosine rule to solve more complex problems, involving right-angled and non right-angled triangles
- [ ] solve 3-D problems using trigonometry
- [ ] find two angles between 0° and 360° for values of sine and cosine
- [ ] solve simple trigonometric equations where sine or cosine is the subject
- [ ] prove geometrical results with a logical and rigorous argument

You are working at (Grade A*) level.

Basic algebra

D-C

1 a Work out the value of $2a^2 - 3b$, when $a = 3$ and $b = -2$.

_____ **[1 mark]**

b Work out the value of $\dfrac{x^2 - y^2}{z}$, when $x = -2$, $y = -6$, and $z = -8$.

_____ **[2 marks]**

D

2 a Expand $5(x - 3)$.

_____ **[1 mark]**

b Expand and simplify $2(x + 1) + 2(3x + 2)$.

_____ **[2 marks]**

c Expand and simplify $3(x - 4) + 2(4x + 1)$.

_____ **[2 marks]**

d A rectangle has a length of $2x + 3$ and a width of $x + 3$.

Write down and simplify an expression for the perimeter in terms of x.

_____ **[2 marks]**

D

3 Factorise these expressions.

a $4x + 6$

_____ **[1 mark]**

b $5x^2 + 2x$

_____ **[1 mark]**

This page tests you on • substitution • expansion and simplification • factorisation

Linear equations

1 Solve these equations.

a $3x + 4 = 1$

$x =$ _____ **[2 marks]**

b $\dfrac{7x - 2}{3} = 4$

$x =$ _____ **[2 marks]**

c $4(x - 1) = 6$

$x =$ _____ **[2 marks]**

2 Solve these equations.

a $4(3y - 2) = 16$

$y =$ _____ **[2 marks]**

b $5x - 2 = x + 10$

$x =$ _____ **[2 marks]**

c $5x - 2 = 3x + 1$

$x =$ _____ **[2 marks]**

This page tests you on
- solving linear equations
- solving equations with brackets
- equations with the variable on both sides of the equals sign

C

1 Solve these equations.

a $3(x + 4) = x - 5$

$x =$ _____ **[2 marks]**

b $5(x - 2) = 2(x + 4)$

$x =$ _____ **[2 marks]**

C

2 ABC is a triangle with sides, measured in centimetres, of x, $3x - 1$ and $2x + 5$.

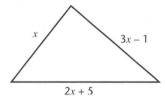

If the perimeter of the triangle is 25 cm, find the value of x.

_____ **[3 marks]**

This page tests you on • equations with brackets and the variable on both sides
• setting up equations

Trial and improvement

C

1 Use trial and improvement to solve the equation $x^3 + 4x = 203$.

The first two entries of the table are filled in.

Complete the table to find the solution. Give your answer to 1 decimal place.

Guess	$x^3 + 4x$	Comment
5	145	Too low
6	240	Too high

$x =$ _____ **[3 marks]**

C

2 Darlene is using trial and improvement to find a solution to this equation.

$$2x + \frac{2}{x} = 8$$

The table shows her first trial.

Complete the table to find the solution. Give your answer to 1 decimal place.

Guess	$2x + \dfrac{2}{x}$	Comment
3	6.66	Too low

$x =$ _____ **[4 marks]**

This page tests you on • trial and improvement

Simultaneous equations

1 Solve these simultaneous equations.

B

    a $5x + 3y = 8$

       $3x - y = 9$

       $x =$ _____

       $y =$ _____ **[2 marks]**

    b $6x + 3y = 9$

       $2x - 3y = 1$

       $x =$ _____

       $y =$ _____ **[2 marks]**

    c $5x + 3y = 6$

       $3x - 7y = 19$

       $x =$ _____

       $y =$ _____ **[2 marks]**

2 a Rearrange the equation $y + 3x = 5$ to make y the subject.

B

       $y =$ _____ **[1 mark]**

    b Solve these simultaneous equations by the method of substitution.

       $y + 3x = 5$

       $2x = 22 - 3y$

       $x =$ _____

       $y =$ _____ **[2 marks]**

This page tests you on   • simultaneous equations

Simultaneous equations and formulae

1 A widget weighs x grams. A whotsit weighs y grams.

24 widgets and 20 whotsits weigh 134 grams.

20 widgets and 24 whotsits weigh 130 grams.

a Write down a pair of simultaneous equations in x and y, using the information above.

_____ **[2 marks]**

b Solve your simultaneous equations and find the weight of a widget.

_____ **[2 marks]**

A

2 Rearrange each of these formulae to make x the subject.

Simplify your answers where possible.

a $C = \pi x$

$x =$ _____ **[1 mark]**

b $6y = 3x - 9$

$x =$ _____ **[2 marks]**

c $3q = 2(4 - x)$

$x =$ _____ **[2 marks]**

d $4y + 3 = 2x + 1$

$x =$ _____ **[2 marks]**

e $4(y + 3) = 2(x + 1)$

$x =$ _____ **[2 marks]**

B-C

This page tests you on
- setting up simultaneous equations
- rearranging formulae

Algebra 2

C

1 a Expand $3(x + 2)$.

_____ **[1 mark]**

b Expand $x(x + 2)$.

_____ **[1 mark]**

c Expand and simplify $(x - 3)(x + 2)$.

_____ **[2 marks]**

d A rectangle has length $x + 2$ and width $x + 1$.

Write down an expression for the area, in terms of x, and simplify it.

$x + 2$

$x + 1$

_____ **[2 marks]**

C

2 a Multiply out and simplify $(x - 4)(x + 1)$.

_____ **[2 marks]**

b Multiply out and simplify $(x + 4)^2$.

_____ **[2 marks]**

E

3 a Work out the value of $3p + 2q$ when $p = -2$ and $q = 5$.

_____ **[1 mark]**

b Find the value of $a^2 + b^2$ when $a = 4$ and $b = 6$.

_____ **[1 mark]**

c An aeroplane has f first-class seats and e economy seats.

For a flight, each first-class seat costs £200 and each economy seat costs £50.

i If all seats are taken, write down an expression in terms of f and e for the total cost of all the seats in the aeroplane.

_____ **[1 mark]**

ii If $f = 20$ and $e = 120$, work out the actual cost of all the seats.

_____ **[1 mark]**

This page tests you on • quadratic expansion • squaring brackets

Factorising quadratic expressions

1 a Factorise $x^2 - 4x$.

_____ **[1 mark]**

b i Factorise $x^2 - 4x - 12$.

_____ **[2 marks]**

ii Solve the equation $x^2 - 4x - 12 = 0$.

_____ **[1 mark]**

B

---

2 a Expand and simplify $(x + y)(x - y)$.

_____ **[1 mark]**

b Factorise $x^2 - 49$.

_____ **[1 mark]**

c Factorise $4a^2 - 9$.

_____ **[2 marks]**

A-B

---

3 a Expand and simplify $(2x + 3)(3x - 1)$.

_____ **[1 mark]**

b Factorise $6x^2 - 17x + 12$.

_____ **[2 marks]**

A

---

4 Solve these equations.

a $x^2 + 3x - 10 = 0$

$x =$ _____

$x =$ _____ **[2 marks]**

b $x^2 + 4x - 5 = 0$

$x =$ _____

$x =$ _____ **[2 marks]**

B

This page tests you on
- factorising quadratics with a unit coefficient of x^2
- difference of two squares
- solving quadratic equations by factorisation

Solving quadratic equations

A

1 a Factorise $12x^2 + 7x + 1$.

_____ **[2 marks]**

b Hence solve $12x^2 + 7x + 1 = 0$.

$x =$ _____

$x =$ _____ **[1 mark]**

A

2 Solve these equations.

a $2x^2 - 9 = 0$

$x =$ _____

$x =$ _____ **[2 marks]**

b $4x^2 - 20x = 0$

$x =$ _____

$x =$ _____ **[2 marks]**

A

3 a Solve the equation $x^2 + 3x - 7 = 0$. Give your answers to 2 decimal places.

$x =$ _____

$x =$ _____ **[2 marks]**

b Solve the equation $2x^2 + 5x - 4 = 0$. Give your answers to 2 decimal places.

$x =$ _____

$x =$ _____ **[2 marks]**

A

4 Solve these equations.

a $9x^2 - 9x + 2 = 0$

$x =$ _____

$x =$ _____ **[2 marks]**

b $8x^2 + 6x - 5 = 0$

$x =$ _____

$x =$ _____ **[2 marks]**

This page tests you on
- solving quadratics of the form $ax^2 + bx + c = 0$
- solving a quadratic equation by factorisation
- solving the general quadratic by the quadratic formula

1 a Find the values of a and b such that

$x^2 + 8x - 7 = (x + a)^2 - b$.

$a = $ _____

$b = $ _____ **[2 marks]**

b Hence solve $x^2 + 8x - 7 = 0$.

Give your answers in surd form.

$x = $ _____

$x = $ _____ **[2 marks]**

A

2 a Find the values of a and b such that

$x^2 + 10x - 3 = (x + a)^2 - b$.

$a = $ _____

$b = $ _____ **[2 marks]**

b Hence solve $x^2 + 10x - 3 = 0$.

Give your answers to 2 decimal places.

$x = $ _____

$x = $ _____ **[2 marks]**

A

3 a Find the values of a and b such that

$x^2 + 2x + 7 = (x + a)^2 + b$.

$a = $ _____

$b = $ _____ **[2 marks]**

b Explain why the equation $x^2 + 2x + 7 = 0$ has no solution.

_____ **[1 mark]**

A

4 a Find the values of a and b such that

$x^2 - 8x + 2 = (x + a)^2 - b$.

$a = $ _____

$b = $ _____ **[2 marks]**

b Hence solve $x^2 - 8x + 2 = 0$.

Give your answers to 2 decimal places.

$x = $ _____

$x = $ _____ **[2 marks]**

A

This page tests you on • solving a quadratic equation by completing the square
• quadratic equations with no solution

Grades

A

1 Use the quadratic formula to solve the equation $x^2 + 5x - 9 = 0$.

Leave your answers in surd form.

$x =$ _____ **[2 marks]**

A*

2 The sides of a right-angled triangle are $(5x + 4)$ cm, $(6x + 1)$ cm and $(2x - 1)$ cm.

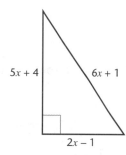

5x + 4   6x + 1

2x − 1

a Explain why $7x^2 - 24x - 16 = 0$.

_____

_____ **[2 marks]**

b Find the value of x.

_____ cm **[2 marks]**

A*

3 A number plus its reciprocal equals 2.9.

Let the number be x.

a Set up a quadratic equation in x.

_____ **[2 marks]**

b Solve the equation to find x.

$x =$ _____ **[2 marks]**

This page tests you on
- **using the quadratic formula without a calculator**
- **solving problems with quadratic equations**

Real-life graphs

1 Martin walked from his house to a viewpoint 5 kilometres away, and back again.

The distance–time graph shows his journey.

a The viewpoint is uphill from Martin's house.

Martin took a rest before walking up the steepest part of the hill.

i How far from home was Martin when he took a rest? _____ km **[1 mark]**

ii For how long did Martin rest? _____ minutes **[1 mark]**

b Martin stopped at the viewpoint before returning home.

He then walked home at a fast, steady pace.

i How long did it take Martin to walk home? _____ minutes **[1 mark]**

ii What was Martin's average speed on the way home? _____ km/hour **[1 mark]**

2 A water tank is empty. It holds a total of 10 000 litres. It is filled at a rate of 2000 litres per minute for 3 minutes and then 1000 litres per minute for 4 minutes.

It then empties at a rate of 2500 litres per minute.

Show this information on these axes.

[2 marks]

This page tests you on • travel graphs • real-life graphs

Trigonometry

A

1 ACD is a right-angled triangle.

B is a point on AC.

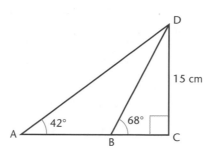

a Calculate the length BC.

_____ cm   **[2 marks]**

b Calculate the length AB.

_____ cm   **[2 marks]**

A

2 O is the centre of a circle of radius 8 cm.

AB and BC are chords of equal length that intersect at B.

AO and OC intersect at 130°.

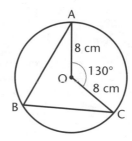

a Use the isosceles triangle OAC to work out the length AC.

_____ cm   **[2 marks]**

b Use circle theorems to write down the size of angle ABC.

_____ °   **[1 mark]**

c Use the isosceles triangle ABC to work out the length BC.

_____ cm   **[2 marks]**

This page tests you on   • 2-D trigonometric problems

3-D trigonometry

1 A cuboid has sides of 6 cm, 8 cm and 15 cm.

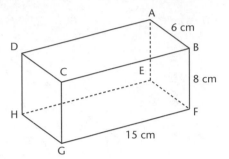

Calculate the size of angle BDF.

_____ ° **[2 marks]**

2 A triangular-based pyramid, ABCD, has an isosceles triangular base ABC, with AC = BC = 12 cm.

One face, ADB, is an equilateral triangle with sides of 8 cm.

D is directly above the mid-point, M, of the side AB.

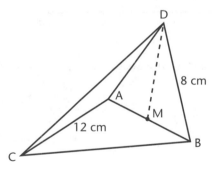

Find the size of angle DCM.

_____ ° **[2 marks]**

This page tests you on • 3-D trigonometric problems

Trigonometric ratios of angles from 0° to 360°

A*

1 The graph shows
$y = \sin x$ for
$0° \leqslant x \leqslant 360°$.

Sine x

a Use the graph to estimate two angles that have a sine of −0.6.

_____ ° and _____ ° **[2 marks]**

b You are told that $\sin^{-1} 0.15 = 8.6°$.

Write down two angles that have a sine of −0.15.

_____ ° and _____ ° **[2 marks]**

A*

2 The graph shows
$y = \cos x$ for
$0° \leqslant x \leqslant 360°$.

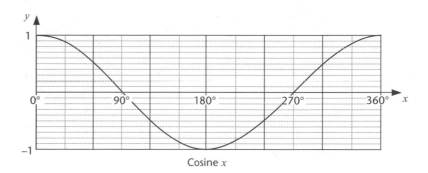

Cosine x

a Use the graph to estimate two angles that have a cosine of 0.75.

_____ ° and _____ ° **[2 marks]**

b You are told that $\cos^{-1} 0.25 = 75.5°$.

Write down two angles that have a cosine of −0.25.

_____ ° and _____ ° **[2 marks]**

This page tests you on • trigonometric ratio of angles from 0° to 360°
• sine and cosine values of angles from 0° to 360°

Sine rule

1 Find the size of angle ABC in this triangle.

_____ ° **[2 marks]**

2 a Find the value of x in this triangle.

_____ cm **[2 marks]**

b In this triangle, angle ABC is obtuse.

Find the size of angle ABC.

_____ ° **[2 marks]**

This page tests you on • solving any triangle • the sine rule
• the ambiguous case

Cosine rule

1 Find the size of angle ACB in this triangle.

_____ ° **[2 marks]**

2 a Find the value of x in this triangle.

_____ cm **[2 marks]**

b In this triangle, angle ABC is obtuse.

Find the size of angle ABC.

_____ ° **[2 marks]**

This page tests you on • the cosine rule

Solving triangles

1 ABCD is a quadrilateral.

a Find the length of the diagonal AC.

_____ cm **[2 marks]**

b Find the size of angle DCA.

_____ ° **[2 marks]**

2 a Work out the area of this triangle.

_____ cm² **[2 marks]**

b In the triangle PQR, PQ = 7 cm, PR = 12 cm and angle PQR = 106°.

Work out the area of PQR.

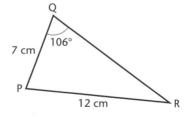

_____ cm² **[2 marks]**

This page tests you on
- which rule to use
- using sine to find the area of a triangle

Linear graphs

1 Draw the graph of $y = 2x - 1$ for $-3 \leqslant x \leqslant 3$ on the axes below.

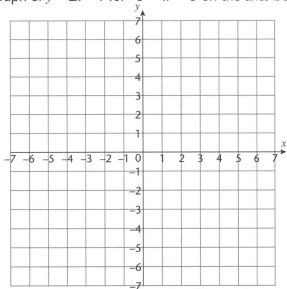

[2 marks]

2 a What is the gradient of each of these lines?

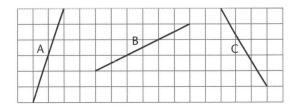

Line A: _____ Line B: _____ Line C: _____ **[3 marks]**

b On the grid below, draw lines with gradients of:

i 2   **ii** $-\frac{1}{2}$ **iii** $\frac{3}{2}$

[3 marks]

This page tests you on   • linear graphs • gradients

1 Here are the equations of six lines.

A $y = 3x + 6$ **B** $y = 2x - 1$ **C** $y = \frac{1}{2}x - 1$

D $y = 3x + 1$ **E** $y = \frac{1}{3}x + 1$ **F** $y = 4x + 2$

a Which line is parallel to line A?

_____ **[1 mark]**

b Which line crosses the y-axis at the same point as line B?

_____ **[1 mark]**

c Which other two lines intersect on the y-axis?

_____ and _____ **[1 mark]**

d Use the gradient-intercept method to draw the graph of $y = 3x - 2$ for $-3 \leq x \leq 3$ on the axes supplied.

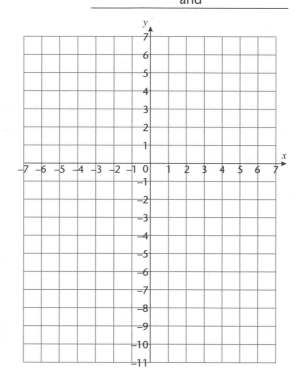

[2 marks]

C

2 Use the cover-up method to draw these graphs on the axes supplied.

a $5x - 2y = 10$

b $2x - 3y = 6$

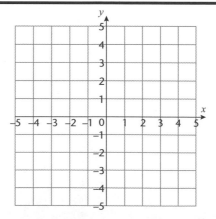

[2 marks]

B

This page tests you on
- **the gradient-intercept method**
- **cover-up method for drawing graphs**

Equations of lines

1 a What is the equation of the line shown?

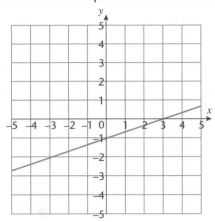

_____ **[2 marks]**

b On the same axes, draw the line $y = 2x + 4$. **[2 marks]**

c Where do the graphs intersect?

_____ **[1 mark]**

2 Frank compares two of his fuel bills.

For the first quarter of the year he used 200 units and paid £35.

For the second quarter of the year he used 150 units and paid £30.

a Plot this information on the axes below.

[2 marks]

b Join the points with a straight line and extend it to meet the axis. **[1 mark]**

c Use the graph to establish a formula between the cost of fuel, C, and the number of units used, n.

_____ **[2 marks]**

This page tests you on
- finding the equation of a line from its graph
- uses of graphs – finding formulae or rules

Linear graphs and equations

1 a Draw the graph of $2x + 3y = 6$
on the axes opposite.

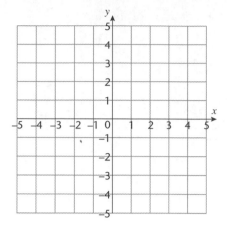

[2 marks]

b On the same axes, draw the line $y = 2x - 2$. **[2 marks]**

c Use the graph to find the solution to these simultaneous equations.
$2x + 3y = 6$
$y = 2x - 2$

$x = $ _____

$y = $ _____ **[1 mark]**

2 A is the point $(-4, -2)$ and
B is the point $(5, 1)$.

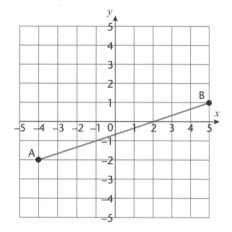

a What is the mid-point of AB?

_____ **[1 mark]**

b What is the gradient of AB?

_____ **[1 mark]**

c Find the equation of the line perpendicular to AB that passes through its
mid-point.

_____ **[2 marks]**

This page tests you on • uses of graphs – solving simultaneous equations
• parallel and perpendicular lines

Quadratic graphs

C

1 a Complete the table of values for $y = x^2 - 2x + 1$.

x	−2	−1	0	1	2	3	4
y	9	4	1				

[2 marks]

b Draw the graph of $y = x^2 - 2x + 1$ for values of x from −2 to 4.

[2 marks]

c Use the graph to find the value(s) of x when $y = 6$.

_____ **[1 mark]**

d Use the graph to solve the equation $x^2 - 2x + 1 = 0$.

$x =$ _____ **[1 mark]**

C

2 a Complete the table of values for $y = x^2 + 2x - 1$.

x	−4	−3	−2	−1	0	1	2
y		2	−1				7

[2 marks]

b Draw the graph of $y = x^2 + 2x - 1$ for values of x from −4 to 2.

[2 marks]

c Use the graph to find the value(s) of x when $y = 1.5$.

_____ **[1 mark]**

d Use the graph to solve the equation $x^2 + 2x - 1 = 0$.

$x =$ _____ **[1 mark]**

This page tests you on
- drawing quadratic graphs
- reading values from quadratic graphs
- using graphs to solve quadratic equations

Non-linear graphs

1 Below is the graph of $y = x^2 - x - 6$.

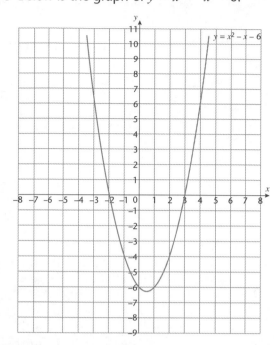

a Use the graph to solve
$x^2 - x - 6 = 0$.

$x = $ _____

$x = $ _____ **[1 mark]**

b Deduce the solutions to the
equation $x^2 - x - 12 = 0$.

$x = $ _____

$x = $ _____ **[1 mark]**

c By drawing a suitable straight
line, solve the equation
$x^2 - 2x - 8 = 0$.

$x = $ _____

$x = $ _____ **[2 marks]**

2 Below is the graph of $y = 12x - x^3$ for values of x from -4 to 4.

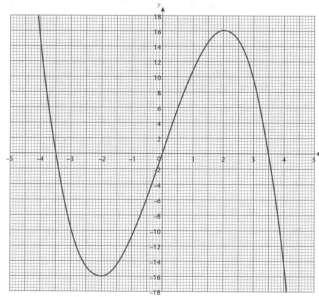

a Use the graph to solve $12x - x^3 = 0$.

_____ **[1 mark]**

b By drawing a suitable straight line, solve the equation $11x - x^3 = 0$.

_____ **[2 marks]**

This page tests you on
- the significant points of a quadratic graph
- solving equations by the method of intersection

Other graphs

A

1 a Complete the table of values for $y = (0.5)^x$.

x	0	1	2	3	4	5
y	1	0.5	0.25			0.03

[2 marks]

b Draw the graph of $y = (0.5)^x$ for values of x from 0 to 5, on the axes opposite.

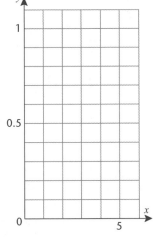

[2 marks]

c Use your graph to solve the equation $0.5^x = 0.8$.

_____ **[1 mark]**

A*

2 The sketch shows the graph of $y = x^3 + 2x^2 - 16x - 32$.

The expression $x^3 + 2x^2 - 16x - 32$ factorises to give $(x + 4)(x - 4)(x + 2)$.

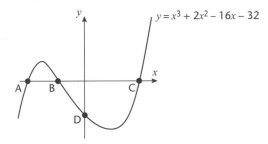

Write down the coordinates of:

A _____ **[1 mark]**

B _____ **[1 mark]**

C _____ **[1 mark]**

D _____ **[1 mark]**

This page tests you on • reciprocal graphs • cubic graphs
• exponential graphs

Algebraic fractions

A

1 a Simplify $\dfrac{2x + 3}{2} + \dfrac{3x - 1}{3}$.

_____ **[2 marks]**

b Simplify $\dfrac{3x - 1}{5} - \dfrac{2x - 5}{2}$.

_____ **[2 marks]**

c Simplify $\dfrac{2x - 1}{3} \times \dfrac{4x - 1}{2}$.

_____ **[2 marks]**

d Simplify $\dfrac{x - 1}{3} \div \dfrac{3x - 1}{6}$.

_____ **[2 marks]**

2 Simplify the following. Factorise and cancel where possible.

A*

a $\dfrac{2x + 3}{5} \div \dfrac{6x + 9}{15}$

_____ **[2 marks]**

b $\dfrac{2x^2}{9} - \dfrac{2y^2}{3}$

_____ **[2 marks]**

This page tests you on   •   algebraic fractions

Solving equations

1 a Solve the equation $\dfrac{x+1}{2} + \dfrac{x-3}{5} = 2$.

$x =$ _____ **[2 marks]**

b Solve the equation $\dfrac{4x+1}{3} - \dfrac{3x-1}{5} = 2$.

$x =$ _____ **[2 marks]**

c Solve the equation $\dfrac{3}{5x+1} + \dfrac{5}{3x-1} = 3$.

$x =$ _____ **[2 marks]**

2 a Show that the equation $\dfrac{3}{2x-1} - \dfrac{4}{3x-1} = 1$ simplifies to $x^2 - x = 0$.

_____ **[2 marks]**

b Hence, or otherwise, solve the equation $\dfrac{3}{2x-1} - \dfrac{4}{3x-1} = 1$.

$x =$ _____ **[1 mark]**

3 Solve this equation.

$$\dfrac{x}{x-1} + \dfrac{3}{x+1} = 1$$

$x =$ _____ **[2 marks]**

This page tests you on • solving equations with algebraic fractions

Simultaneous equations 2

1 Solve these simultaneous equations.

$y = 4 - x$

$x^2 + y = 16$

[3 marks]

2 The diagram shows the line $2y = 3x - 13$ and the curve $x^2 + y^2 = 65$.

The curve and the line intersect at the points A and B.

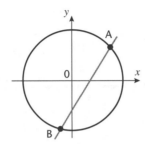

a Rearrange the equation $2y = 3x - 13$ to make y the subject.

[1 mark]

b By solving the simultaneous equations $2y = 3x - 13$ and $x^2 + y^2 = 65$, find the coordinates of A and B.

[3 marks]

3 Solve these simultaneous equations.

$x^2 + 2y^2 = 11$

$x = 3y$

[3 marks]

This page tests you on • linear and non-linear simultaneous equations

The nth term

C

1 The nth term of a sequence is $4n + 1$.

   a Write down the first three terms of the sequence.

      _____ **[1 mark]**

   b Which term of the sequence is equal to *29*?

      _____ **[1 mark]**

   c Explain why 84 is *not* a term in this sequence.

   _____

   _____ **[1 mark]**

C

2 What is the nth term of the sequence 3, 10, 17, 24, 31, …?

      _____ **[2 marks]**

D

3 R is an odd number. Q is an even number. P is a prime number.

   Tick the appropriate box to identify these expressions as *always even, always odd* or *could be either*.

	Always even	Always odd	Could be either	
a $R + Q$	☐	☐	☐	**[1 mark]**
b RQ	☐	☐	☐	**[1 mark]**
c $P + Q$	☐	☐	☐	**[1 mark]**
d R^2	☐	☐	☐	**[1 mark]**
e $R + PQ$	☐	☐	☐	**[1 mark]**

C

4 a n is a positive integer. Explain why $2n$ is always an even number.

   _____

   _____ **[1 mark]**

   b Zoe says that when you square an even number you always get a multiple of 4. Show that Zoe is correct.

   _____

   _____ **[2 marks]**

This page tests you on   • nth term of a sequence   • special sequences

Formulae

1 Matches are used to make patterns with hexagons.

Pattern 1   **Pattern 2**   **Pattern 3**   **Pattern 4**

a Complete the table that shows the number of matches used to make each pattern.

Pattern number	1	2	3	4	5
Number of matches	6	11			

b How many matches will be needed to make the 20th pattern?

_____ **[1 mark]**

c How many matches will be needed to make the nth pattern?

_____ **[2 marks]**

C

2 Rearrange the formula to make x the subject.

$$y = \frac{x + 2}{x - 4}$$

_____ **[2 marks]**

A

3 Rearrange the formula to make x the subject.

Simplify your answer as much as possible.

$$6x + 2y = 4(x + 3)$$

_____ **[3 marks]**

A

This page tests you on
- the nth term from given patterns
- changing the subject of a formula

Inequalities

1 a Solve the inequality $3x - 4 \leqslant 2$.

_____ **[1 mark]**

b What inequality is shown on this number line?

_____ **[1 mark]**

c Write down all the integers that satisfy both inequalities in parts **a** and **b**.

_____ **[1 mark]**

2 a What inequality is shown on this number line?

_____ **[1 mark]**

b Solve these inequalities.

i $\dfrac{x}{2} + 3 > 1$

_____ **[2 marks]**

ii $\dfrac{(x + 3)}{2} \leqslant 1$

_____

c Write down all the integers that satisfy both inequalities in parts **a** and **b**.

_____ **[2 marks]**

3 Solve these inequalities.

a $3x - 2 \geqslant x + 7$

_____ **[2 marks]**

b $3(x - 1) < x - 5$

_____ **[2 marks]**

This page tests you on • solving inequalities • inequalities on number lines

Graphical inequalities

1 On the graph, the region **R** is shaded.   **[1 mark]**

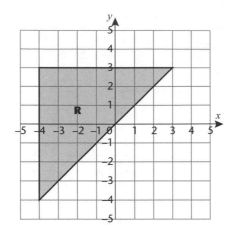

Write down the three inequalities that together describe the shaded region.

_____

_____

_____ **[3 marks]**

2 On the grid below, indicate the region defined by these three inequalities.

Mark the region clearly as **R**.

$x \leqslant 2$

$x + y > -3$

$y > -3$

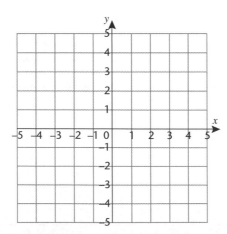

[3 marks]

This page tests you on • graphical inequalities • more than one inequality

Graph transforms

1 The graph of $y = \sin x$ is shown by a dotted line on the axes in parts **a** to **c** below.

Sketch the graphs stated in each case.

a $y = \sin x - 1$

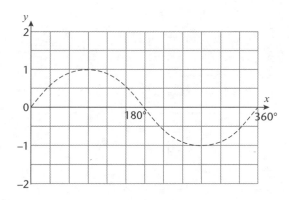

[1 mark]

b $y = \frac{1}{2} \sin x$

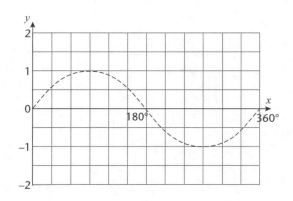

[1 mark]

c $y = \sin (x + 90°)$

[1 mark]

1 The graph of $y = \sin x$ is shown by a dotted line on the axes in parts **a** to **c** below.

Sketch the graphs stated in each case.

a $y = \sin 3x$

[1 mark]

b $y = -\sin x$

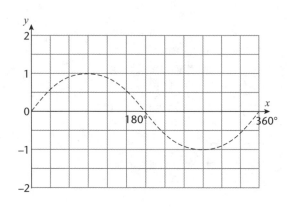

[1 mark]

c $y = -\sin x + 1$

[1 mark]

This page tests you on • transformations of the graph $y = f(x)$

Proof

A*

1 Prove that $(n + 5) - (n + 3)^2 = 4(n + 4)$.

[3 marks]

A*

2 ABCD is a cyclic quadrilateral.

Triangle ADB is an isosceles triangle where AB = AD.

Prove that AD is parallel to CB.

Give reasons for any angle values that you calculate or write down.

[3 marks]

A*

3 The nth term of the triangular number sequence is given by $\frac{1}{2}n(n + 1)$.

Let any term in the triangular number sequence be T.

Prove that $8T + 1$ is always a square number.

[3 marks]

Algebra checklist

I can...

- [ ] use letters to write more complicated algebraic expressions
- [ ] expand expressions with brackets
- [ ] factorise simple expressions
- [ ] solve linear equations where the variable appears on both sides of the equals sign
- [ ] solve linear equations that require the expansion of a bracket
- [ ] set up and solve simple equations from real-life situations
- [ ] find the average speed from a travel graph
- [ ] draw a linear graph without a table of values
- [ ] substitute numbers into an nth-term rule
- [ ] understand how odd and even numbers interact in addition, subtraction and multiplication problems

You are working at (Grade D) level.

- [ ] expand and simplify expressions involving brackets
- [ ] factorise expressions involving letters and numbers
- [ ] expand pairs of linear brackets to give a quadratic expression
- [ ] solve linear equations that have the variable on both sides and include brackets
- [ ] solve simple linear inequalities
- [ ] show inequalities on a number line
- [ ] solve equations, using trial and improvement
- [ ] rearrange simple formulae
- [ ] give the nth term of a linear sequence
- [ ] give the nth term of a sequence of powers of 2 or 10
- [ ] draw a quadratic graph, using a table of values

You are working at (Grade C) level.

- [ ] solve a quadratic equation from a graph
- [ ] recognise the shape of graphs $y = x^3$ and $y = \dfrac{1}{x}$
- [ ] solve two linear simultaneous equations
- [ ] rearrange more complex formula to make another variable the subject
- [ ] factorise a quadratic expression of the form $x^2 + ax + b$
- [ ] solve a quadratic equation of the form $x^2 + ax + b = 0$
- [ ] interpret real-life graphs
- [ ] find the equation of a given linear graph
- [ ] solve a pair of linear simultaneous equations from their graphs
- [ ] draw cubic graphs, using a table of values
- [ ] use the nth term to generate a quadratic sequence

- [ ] solve equations involving algebraic fractions where the subject appears as the numerator
- [ ] solve more complex linear inequalities
- [ ] represent a graphical inequality on a coordinate grid
- [ ] find the inequality represented by a graphical inequality
- [ ] verify results by substituting numbers

You are working at (Grade B) level.

- [ ] draw exponential and reciprocal graphs, using a table of values
- [ ] find the proportionality equation from a direct or inverse proportion problem
- [ ] set up and solve two simultaneous equations from a practical problem
- [ ] factorise a quadratic expression of the form $ax^2 + bx + c$
- [ ] solve a quadratic equation of the form $ax^2 + bx + c = 0$ by factorisation
- [ ] solve a quadratic equation of the form $ax^2 + bx + c = 0$ by the quadratic formula
- [ ] write a quadratic expression of the form $x^2 + ax + b$ in the form $(x + p)^2 + q$
- [ ] interpret and draw more complex real-life graphs
- [ ] find the equations of graphs parallel and perpendicular to other lines and passing through specific points
- [ ] rearrange a formula where the subject appears twice
- [ ] combine algebraic fractions, using the four rules of addition, subtraction, multiplication and division
- [ ] translate and solve a real-life problem, using inequalities
- [ ] show that a statement is true, using verbal or mathematical arguments

You are working at (Grade A) level.

- [ ] solve equations using the intersection of two graphs
- [ ] use trigonometric graphs to solve sine and cosine problems
- [ ] solve direct and inverse proportion problems, using three variables
- [ ] solve a quadratic equation of the form $x^2 + ax + b = 0$ by completing the square
- [ ] solve real-life problems that lead to a quadratic equation
- [ ] solve quadratic equations involving algebraic fractions where the subject appears as the denominator
- [ ] rearrange more complicated formulae where the subject may appear twice
- [ ] simplify algebraic fractions by factorisation and cancellation
- [ ] solve a pair of simultaneous equations where one is linear and one is non-linear
- [ ] transform the graph of a given function
- [ ] identify the equation of a transformed graph
- [ ] prove algebraic results with rigorous and logical arguments

You are working at (Grade A*) level.

Formulae sheet

Area of trapezium $= \frac{1}{2}(a + b)h$

Volume of prism = acrea of cross-section $\times$ length

Volume of a sphere $= \frac{4}{3}\pi r^3$

Surface area of a sphere $= 4\pi r^2$

Volume of a cone $= \frac{1}{3}\pi r^2 h$

Curved surface area of a cone $= \pi r l$

In any triangle: ABC

Area of triangle $= \dfrac{1}{2} ab \sin C$

Sine rule $\dfrac{a}{\sin A} = \dfrac{b}{\sin B} = \dfrac{c}{\sin C}$

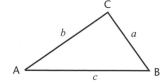

Cosine rule $a^2 = b^2 + c^2 - 2bc \cos A$

The quadratic equation

The solutions for $ax^2 + bx + c = 0$, where $a \neq 0$, are given by

$$x = \frac{-b \pm \sqrt{(b^2 - 4ac)}}{2a}$$

Notes

Notes

Notes

Notes

Answers

Handling data
Page 100 Statistics 1

1 a 2

b 2

c 1.8   *(1 mark for total of 180)*

2 a 0

b 1

c 1.4   *(1 mark for 70)*

d There are bigger numbers in top half of table

Page 101 Statistics 1

1 a 65   *(1 mark for total 13000)*

b

c Girls did better; polygons are about the same shape and girls are about 10 marks better

2 a 1.5, 4.8, 7.2, 1.8, 0.35

b

Page 102 Statistics 1

1 a Because there are 4 quarters and seasons in a year

b 188

c

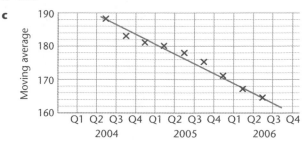

d 190   *(1 mark for 4 × 162 – (158 + 120 + 180)*

2 a

	Boys	Girls
0 < time (hours) ⩽ 4		
4 < time (hours) ⩽ 6		
6 < time (hours) ⩽ 8		
8 < time (hours) ⩽ 10		
More than 10 hours		

(1 mark for times or for boys/girls)

b Not really; the difference is not that large and the sample of girls was too small

Page 103 Statistics 1

1 a 94.8p

b 2006; price index shot up

c Faster; petrol has increased by 58%

2 a Any from: sample too small; sample not representative; sample not random; conclusion incorrect, rounds to 6 out of 10

b Question not specific

Page 104 Statistics 2

1 a 10am–11am

b 11am

c 25 °C

d No; anything could happen in 4 hours it is too far from the end of the graph

2 a 30

b 9.6   *(1 mark for 288)*

Page 105 Scatter diagrams

1 a

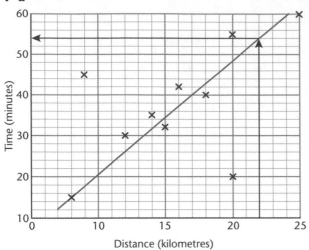

(1 mark for 8 or 9 values correctly plotted)

b i G

  ii I

c *(See graph)*

d 54 minutes

Page 106 Cumulative frequency and box plots

1 a $3\frac{1}{4}$ minutes (3 m 15 s)

  b 1.4 minutes (1 m 24 s)

  c 48

2 a 12

  b 52

  c 10K higher median; highest score

  d 10J smaller interquartile range; smaller range

Page 107 Probability

1 a 0.15, 0.06, 0.18, 0.18, 0.29, 0.14

  b 5; as score 5 had the most frequent result

2 a Because they cannot happen at the same time

  b Because their probabilities add up to 1

  c The probability of red is $\frac{2}{5}$

  d 12

  e 80

Page 108 Probability

1 a 22

  b PE

  c Maths $\frac{5}{12} \approx 42\%$; Science $\frac{7}{18} \approx 39\%$

  d $\frac{12}{40} = \frac{3}{10}$

2 a 0.24

  b 1200

  c 6240

Page 109 Probability

1 a

		First score			
		1	2	3	4
	1	2	2	4	4
Second	2	2	4	6	8
score	3	4	6	6	12
	4	4	8	12	16

  b 1

  c $\frac{6}{16} = \frac{3}{8}$

2 a $\frac{7}{10}, \frac{3}{10}, \frac{7}{10}$

  b $\frac{9}{100}$

  c $\frac{42}{100} = \frac{21}{50}$

Page 110 Probability

1 a $\frac{1}{5}$

  b 24

  c 18

2 a $1 - x$

  b x^3

3 a $\frac{4}{6}, \frac{1}{5}, \frac{4}{5}, \frac{2}{5}, \frac{3}{5}$

  b $\frac{14}{30} = \frac{7}{15}$

Number
Page 112 Number

1 a 572   *(1 mark for 476 and 96)*

  b 7 of each   *(1 mark for 128 ÷ 20)*

2 a i 18

    ii 24

  b i £6.24

    ii £300

Page 113 Number

1 a i 70 000

   ii 0.067

  b i 6000

   ii 250

2 a i 370

   ii 250

  b i 0.76

   ii 0.0065

  c i 12 000 000

   ii 360 000

  d i 3 000

   ii 500

Page 114 Number

1 a $p = 2$, $q = 3$

  b $2^3 \times 3^2 \times 5$

  c $a = 2$, $b = 7$

  d $2^2 \times 7^2$

2 a $2^3 \times 3$

  b $2^2 \times 3 \times 5$

  c 120

  d 12

  e i $2^4 \times 3^2 \times 5^2$

   ii $2^2 \times 3 \times 5$

Page 115 Fractions

1 a $1\frac{3}{20}$ *(1 mark for $\frac{15}{20} + \frac{8}{20}$)*

  b $1\frac{13}{20}$ *(1 mark for $\frac{11}{3} - \frac{9}{5}$)*

  c $\frac{11}{60}$ *(1 mark for $\frac{49}{60}$)*

2 a $3\frac{1}{2}$ *(1 mark for $\frac{5}{2} \times \frac{7}{5}$)*

  b $1\frac{3}{8}$ *(1 mark for $\frac{10}{33} \times \frac{5}{12}$)*

  c $\frac{39}{40}$ *(1 mark for $\frac{1}{2} \times \frac{13}{6} \times \frac{9}{10}$)*

Page 116 Percentage

1 a $6000 \times 0.88 \times 0.9$

  b £6.80 + £3.40 + £1.70 *(1 mark)*; £11.90

2 a £822.50   *(1 mark for 1.175×700)*

  b £220   *(1 mark for 0.88×250)*

3 18%   *(1 mark for $405 \div 2250$)*

Page 117 Percentage

1 a $8000 \times 0.88 \times 0.88$

  b £2458.51   *(1 mark for 2000×1.035^6)*

  c £1771.68   *(1 mark for 2000×0.98^6)*

2 a Electro £228, Corries £228.80

  b £380   *(1 mark for $361 \div 0.95$)*

Page 118 Ratio

1 a 4 : 3

  b 1 : 0.4

  c 375 ml

2 35   *(1 mark for $15 \div 3 \; (= 5)$)*

3 a 32 km per hour

       (1 mark for $72 \div 2.25$; 1 mark for units)

  b 36 km per hour

       (Deduct a mark if units not included)

Page 119 Ratio

1 a 30   *(1 mark for 5.5)*

  b 143 miles

2 Handy size 3.87 g/p compared to large size 3.63 g/p   *(1 mark for grams ÷ pence)*

3 0.0068 kg/cm³ or 6.8 g/cm²   *(1 mark for units)*

Page 120 Powers and reciprocals

1 a i 13

   ii 125

   b i 4

   ii 256

2 a 64, 4; 256, 6; 1024, 4

   b 4; all odd powers end in 4

   c $5^6 = 15625 > 6^5 = 7776$

3 a $\frac{1}{8}$

   b $\frac{1}{16}$

   c $\frac{3}{8}$ *(1 mark for $\frac{1}{8} + \frac{1}{4}$)*

Page 121 Powers and reciprocals

1 a 4.52×10^4

   b 0.006

   c 1.8×10^6

   d 8×10^{-4}

   e 2×10^2

   f 6×10^2

2 a i 0.4444...

   ii 0.5555...

   b i $\frac{1}{10}$

   ii $1\frac{1}{3}$

   c i $0.\dot{1}$

   ii $1.\dot{3}$

   d i $0.\dot{8}$

   ii $0.\dot{4}$

   iii $0.\dot{2}$

Page 122 Powers and reciprocals

1 a x^7

   b x^4

   c i 4

   ii 3

   d 1 000 000

2 a $16a^4b^3$ *(1 mark for top line as $32a^5b^5$)*

   b $9x^4y^6$ *(1 mark for any two correct terms)*

3 a 2

   b $\frac{1}{11}$

   c $\frac{1}{27}$ *(1 mark for 27)*

Page 123 Surds

1 a $4\sqrt{3}$

   b $9\sqrt{3}$

   c $-2 + 4\sqrt{3}$ *(1 mark for each term)*

   d 1   *(1 mark for $5 + 2\sqrt{5} - 2\sqrt{5} - 4$)*

2 a $\frac{\sqrt{6}}{2}$ *(1 mark for $\frac{3\sqrt{6}}{6}$)*

   b $2\sqrt{2}$ cm

   c $\frac{(16 + 12)}{4\sqrt{12}}$ *(1 mark)*; $\frac{(28 \times \sqrt{12})}{48}$ *(1 mark)*

Page 124 Variation

1 a $m = ks^3$

   b $\frac{1}{20}$

   c 400

   d 4

2 a i $\frac{1}{8}$

   ii 4

   b 4   *(1 mark for $y = \frac{20}{\sqrt[3]{x}}$)*

Page 125 Limits

1 a i 45

   ii 54

   b i 45

   ii 55

2 a 7414.875 cm³   *(1 mark for 19.5)*

   b 8615.125 cm³   *(1 mark for 20.5)*

3 $x =$ 225 to 235

   $y =$ 395 to 405   *(1 mark)*

   $\frac{235}{395^2}$ *(1 mark)*

   $=$ 0.001506   *(1 mark)*

Shape, space and measures
Page 127 Circle and area

1 a 452.4 cm²

   b 62.8 cm

2 50π cm²

3 152 cm²   *(1 mark for $\frac{1}{2} \times (22 + 16) \times 8$)*

Page 128 Sectors and prisms

1 a 20π cm^2 *(1 mark for $100\pi \div 5$)*

  b $16\pi + 20$ cm *(1 mark for $20\pi \div 5 \times 4$)*

2 a 14 cm^2

  b 168 cm^3 *(1 mark for 14×12)*

Page 129 Cylinders and pyramids

1 a 160π cm^3 *(1 mark for $\pi \times 4^2 \times 10$)*

  b 112π cm^2 *(1 mark for $2\pi \times 4^2 + 2 \times \pi \times 4 \times 10$)*

2 Volume whole pyramid = 320 cm^3 *(1 mark)*

  Volume cut off pyramid = 5 cm^3 *(1 mark)*

  Volume frustum = 315 cm^3 *(1 mark)*

Page 130 Cones and spheres

1 Volume 6 cm cone = 1.5707 cm^3 *(1 mark)*

  Volume 3 cm cone = 0.7854 cm^3 *(1 mark)*

  Total volume = 2.356 cm^3 *(1 mark)*

  Mass = 3.5 g *(1 mark)*

2 Volume cylinder = 678.6 cm^3 *(1 mark)*

  Volume hemisphere = 261.8 cm^3 *(1 mark)*

  Volume ashtray = 416.8 cm^3 *(1 mark)*

Page 131 Pythagoras' theorem

1 12.8 cm *(1 mark for $\sqrt{164}$)*

2 10.2 cm *(1 mark for $\sqrt{104}$)*

3 4.5 m *(1 mark for $\sqrt{20.41}$)*

Page 132 Pythagoras' theorem

1 40.2 cm^2 *(1 mark for $\sqrt{131.75}$)*

2 a 14.14 cm *(1 mark for $\sqrt{200}$)*

  b 13.2 cm *(1 mark for $\sqrt{175}$)*

  c 14.14 cm *(1 mark for $\sqrt{200}$)*

Page 133 Using Pythagoras and trigonometry

1 a 0.5449883506

  b 0.54 or 0.545

2 a 8.36 cm *(1 mark for $13 \times \sin 40$)*

  b 32.2° *(1 mark for $\sin^{-1}(8 \div 15)$)*

3 a 22.4 cm *(1 mark for $15 \div \cos 48$)*

  b 63.2° *(1 mark for $\cos^{-1}(9 \div 20)$)*

4 a 3.64 cm *(1 mark for $10 \times \tan 20$)*

  b 31.0° *(1 mark for $\tan^{-1}(3 \div 5)$)*

Page 134 Using Pythagoras and trigonometry

1 44.4° *(1 mark for $\cos^{-1}(20 \div 28)$)*

2 25.7 cm *(1 mark for $12 \div \tan 25$)*

3 11.8 cm *(1 mark for $20 \times \sin 36$)*

4 53.1° *(1 mark for $\sin^{-1}(40 \div 50)$)*

Page 135 Geometry

1 40° *(1 mark for BCD = 100°; 1 mark for DCE = 80°; 1 mark for CDE = 60°)*

2 a Pentagon can be split into 3 triangles *(1 mark)*

    $3 \times 180 = 540$ *(1 mark)*

  b 10 *(1 mark for $360 \div 36$)*

  c Exterior angle = 20 *(1 mark)*; $360 \div 20 = 18$ *(1 mark)*

Page 136 Circle theorems

1 a 27° *(1 mark)* ACB = 90°; angles in triangle *(1 mark)*

  b OC = OB, so isosceles triangle

2 a 102° *(1 mark)*;

    opposite angles in a cyclic quad *(1 mark)*

  b 156° *(1 mark)*;

    angle at centre twice angle at circumference *(1 mark)*

Page 137 Circle theorems

1 a 33° *(1 mark)* AQB = 90°; angles in triangle *(1 mark)*

  b 33° *(1 mark)*; alternate segment *(1 mark)*

2 ACB = x; isosceles *(1 mark)*

  BAP = CAT = x Alternate segment *(1 mark)*

  BAC is symmetric so AM must be at right angles to PT
  (1 mark)

Page 138 Transformation geometry

1 R and S

2 a Translation *(1 mark)* of (–6, –3) *(1 mark)*

  b Triangle at (3, 6), (3, 9), (5, 6)

  c (–7, 5)

Page 139 Transformation geometry

1

2 a, b

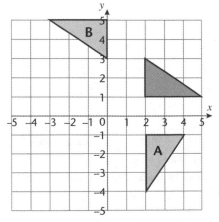

  c 90° *(1 mark)*; clockwise *(1 mark)*;
  about (4, 3) *(1 mark)*

Page 140 Transformation geometry

1 a Enlargement scale factor $\frac{1}{2}$ *(1 mark)*;
  centre (1, 8) *(1 mark)*

  b Triangle at (2, 1), (3, 1), (2, $2\frac{1}{2}$)

2 a, b

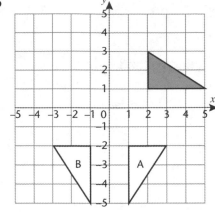

  c Reflection in $y = -x$

Page 141 Constructions

1

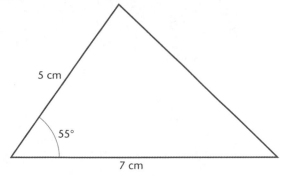

5 cm

55°

7 cm

(1 mark for each length ±1 mm and 1 mark for angle ±1°)

2 a

A

b i

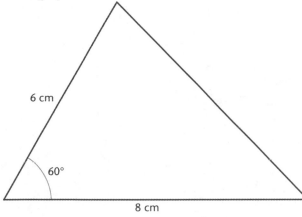

6 cm

60°

8 cm

(1 mark for each length ±1 mm and 1 mark for angle ±1)

ii 7.2 cm

Page 142 Constructions

1 a

A ●

b

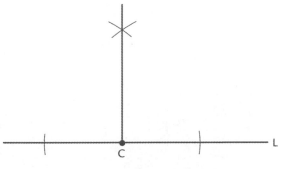

L

C

(Deduct a mark if arcs not shown)

2 a

L

b

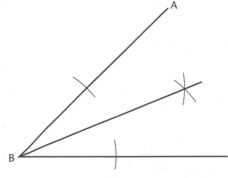

A

B

(Deduct a mark if arcs not shown)

Page 143 Constructions and loci

1

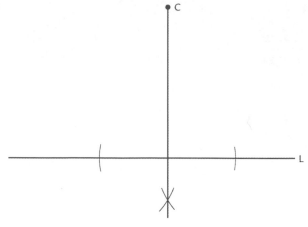

(Deduct a mark if arcs not shown)

2 a, b

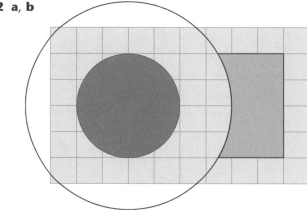

Page 144 Similarity

1 a 8.4 cm

   b 8.8 cm

2 9 m by 6 m *(1 mark for 1500 ÷ 20 or scale factor of 75)*

Page 145 Similarity

1 a 2.7 m × 1.35 m

   b 29.25 m² *(1 mark for $\sqrt{0.75^2}$)*

   c 31.6 cm³ *(1 mark for $\sqrt{0.75^3}$)*

2 a 28.5 cm *(1 mark for $\sqrt[3]{2.5}$)*

   b 38 cm² *(1 mark for $70 ÷ (\sqrt[3]{2.5})^2$)*

Page 146 Dimensional analysis

1 Area; volume; length

2 a Area; perimeter; volume

   b Inconsistent; area plus volume

Page 147 Vectors

1 a

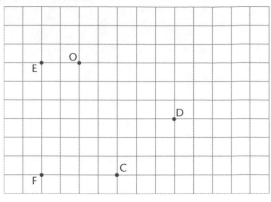

   b $2\frac{1}{2}$**a** + **b**

   c 5**a** − 7**b**

Page 148 Vectors

1 a i $-\frac{1}{2}$**a** + $\frac{1}{2}$**c**

   ii **b** − **a**

   iii **c** − **b**

   iv $-\frac{1}{2}$**a** + $\frac{1}{2}$**c**

   b Parallelogram as PS and QR are equal so parallel and same length

Algebra
Page 151 Basic algebra

1 a 24

   b 4

2 a $5x - 15$

   b $8x + 6$

   c $3x - 12 + 8x + 2$ *(1 mark)* = $11x - 10$

   d $2(2x + 3) + 2(x + 3)$ *(1 mark)* = $6x + 12$

3 a $2(2x + 3)$

   b $x(5x + 2)$

Page 152 Linear equations

1 a $3x = -3$ *(1 mark)* $x = -1$ *(1 mark)*

   b $7x = 14$ *(1 mark)* $x = 2$ *(1 mark)*

   c $4x = 10$ *(1 mark)* $x = 2\frac{1}{2}$ *(1 mark)*

2 a $12y = 24$ *(1 mark)* $y = 2$ *(1 mark)*

   b $4x = 12$ *(1 mark)* $x = 3$ *(1 mark)*

   c $2x = 3$ *(1 mark)* $x = 1\frac{1}{2}$ *(1 mark)*

Page 153 Linear equations

1 a $2x = -17$ *(1 mark)* $x = -8\frac{1}{2}$ *(1 mark)*

   b $3x = 18$ *(1 mark)* $x = 6$ *(1 mark)*

2 $x + 3x - 1 + 2x + 5 = 25$ *(1 mark)*

   $6x + 4 = 25$ *(1 mark)*

   $x = 3.5$ cm *(1 mark)*

Page 154 Trial and improvement

1 *1 mark for finding the answer is between 5.6 and 5.7*

   1 mark for testing 5.65

   1 mark for $x = 5.7$

2 *1 mark for testing 4 (8.5)*

   1 mark for finding the answer is between 3.7 and 3.8

   1 mark for testing 3.75

   1 mark for $x = 3.7$

Page 155 Simultaneous equations

1 a $x = 2\frac{1}{2}; y = -1\frac{1}{2}$

   b $x = 1\frac{1}{4}; y = \frac{1}{2}$

   c $x = 2.25; y = -1.75$

2 a $y = 5 - 3x$

   b $x = -1; y = 8$

Page 156 Simultaneous equations and formulae

1 a $24x + 20y = 134; 20x + 24y = 130$

   b 3.5 grams

2 a $x = \frac{C}{\pi}$

   b $3x = 6y + 9$ *(1 mark)*

       $x = 2y + 3$ *(1 mark)*

   c $x = \frac{(8 - 3q)}{2}$

   d $x = 2y + 1$ *(1 mark for $x = (4y + 2) \div 2$)*

   e $x = 2y + 5$ *(1 mark for $2y + 6 = x + 1$)*

Page 157 Algebra 2

1 a $3x + 6$

   b $x^2 + 2x$

   c $x^2 - 3x + 2x - 6$ *(1 mark)* $= x^2 - x - 6$ *(1 mark)*

   d $(x + 2)(x + 1)$ *(1 mark)* $= x^2 + 3x + 2$ *(1 mark)*

2 a $x^2 - 4x + x - 4$ *(1 mark)* $= x^2 - 3x - 4$ *(1 mark)*

   b $(x + 4)(x + 4)$ *(1 mark)* $= x^2 + 8x + 16$ *(1 mark)*

3 a 4

   b 52

   c i $200f + 50e$

      ii £10 000

Page 158 Factorising quadratic expressions

1 a $x(x - 4)$

   b i $(x - 6)(x + 2)$ *(1 mark each bracket)*

      ii 6, −2

2 a $x^2 - y^2$

   b $(x - 7)(x + 7)$

   c $(2a - 3)(2a + 3)$

3 a $6x^2 + 7x - 3$

   b $(3x - 4)(2x - 3)$

4 a $(x - 2)(x + 5) = 0$ *(1 mark)*

       $x = 2$ or 5 *(1 mark)*

   b $(x - 1)(x + 5) = 0$ *(1 mark)*

       $x = 1$ or $x = -5$ *(1 mark)*

Page 159 Solving quadratic equations

1 a $(3x + 1)(4x + 1)$

   b $x = -\frac{1}{3}; x = -\frac{1}{4}$

2 a $x = \pm\sqrt{4.5} = \pm2.12$

   b $x = 0; x = 5$

3 a $x = 1.54; x = -4.54$

   b $x = 0.64; x = -3.14$

4 a $(3x - 1)(3x - 2) = 0$ *(1 mark)*

       $x = \frac{1}{3}$ or $x = \frac{2}{3}$ *(1 mark)*

   b $(2x - 1)(4x + 5) = 0$ *(1 mark)*

       $x = \frac{1}{2}$ or $x = -\frac{5}{4}$ *(1 mark)*

Page 160 Solving quadratic equations

1 a $a = 4$; $b = -23$

   b $x = -4 \pm \sqrt{23}$

2 a $a = 5$; $b = -28$

   b $x = 0.29$; $x = -10.29$

3 a $a = 1$; $b = 5$

   b Cannot have the square root of a negative

4 a $a = -4$; $b = 14$

   b $x = 7.74$; $x = 0.26$

Page 161 Solving quadratic equations

1 $x = -5 \pm \frac{\sqrt{61}}{2}$

2 a $(6x + 1)^2 = (5x + 4)^2 + (2x - 1)^2$

   b $x = 4$

3 a $x + \frac{1}{x} = 2.9$; $10x^2 - 29x + 10 = 0$

   b $x = \frac{2}{5}$ or $\frac{5}{2}$

Page 162 Real-life graphs

1 a i 2 km

    ii 5 minutes

   b i 40 minutes

    ii $7\frac{1}{2}$ km/h

2

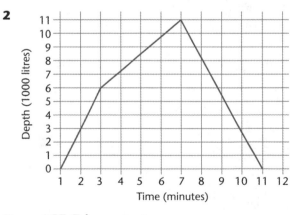

Page 163 Trigonometry

1 a 6.1 cm   *(1 mark for 15 ÷ tan 68)*

   b 10.6 cm   *(1 mark for AC = 16.7)*

2 a 14.5 cm   *(1 mark for 8 × cos 25)*

   b 65°

   c 13.5 cm   *(1 mark for 7.25 ÷ sin 32.5)*

Page 164 3-D trigonometry

1 26.3°   *(1 mark for tan⁻¹ (8 ÷ 16.15))*

2 31.4°   *(1 mark for tan⁻¹ (6.9 ÷ 11.3))*

Page 165 Trigonometric ratios of angles from 0° to 360°

1 a 216.9° and 323.1°

   b 188.6° and 351.4°

2 a 41.4° and 318.6°

   b 104.5° and 255.5°

Page 166 Sine rule

1 51.5°   *(1 mark for sin⁻¹ 0.7825)*

2 a 17cm   *(1 mark for 15 × sin 68 ÷ sin 55)*

   b 127.4°   *(1 mark for 52.6°)*

Page 167 Cosine rule

1 81.2°   *(1 mark for cos⁻¹ 0.153)*

2 a 26.4 cm   *(1 mark for x^2 = 697.6)*

   b 106.2°   *(1 mark for cos⁻¹ −0.279)*

Page 168 Solving triangles

1 a 11.7 cm   *(1 mark for x^2 = 136.8°)*

   b 58.7°   *(1 mark for sin⁻¹ 0.8545)*

2 a 23.1 cm²   *(1 mark for 0.5 × 8 × 9 × sin 40)*

   b 26.9 cm²   *(1 mark for angle R = 34.1°)*

Page 169 Linear graphs

1

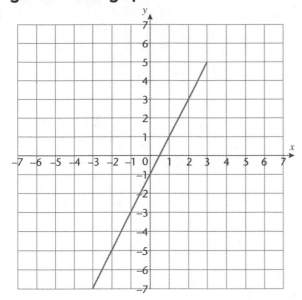

2 a Line A: 3; line B: $\frac{1}{2}$; line C: $-\frac{5}{3}$

b

Page 170 Linear graphs

1 a D

  b C

  c D and E

  d *Graph intercepting y-axis at –2 (1 mark); gradient 3 (1 mark)*

2 a Graph from (2, 0) to (0, –5)

  b Graph from (3, 0) to (0, –2)

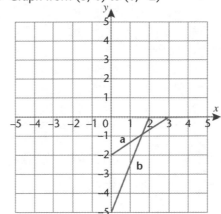

Page 171 Equations of lines

1 a $y = \frac{1}{3}x - 1$

b

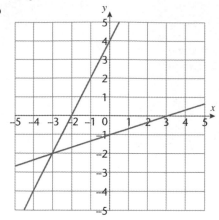

c (−3, −2)

2 a, b

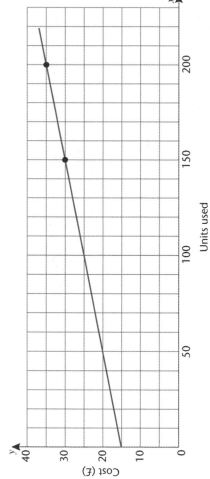

Cost (£)

Units used

c $C = 15 + \frac{1}{10}n$

Page 172 Linear graphs and equations

1 a, b

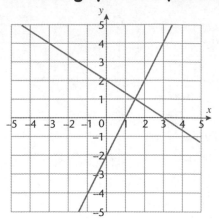

c $x = 1\frac{1}{2}$, $y = 1$

2 a $\left(\frac{1}{2}, -\frac{1}{2}\right)$

b $\frac{1}{3}$

c $y = -3x + 1$

Page 173 Quadratic graphs

1 a 0, 1, 4, 9

b

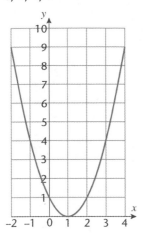

c −1.45, 3.45

d $x = 1$

2 a 7, −2, −1, 2

b

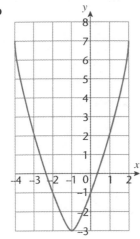

c −2.9, 0.9

d $x = -2.4$ and $x = 0.4$

Page 174 Non-linear graphs

1 a $x = -2; x = 3$

b $x = -3; x = 4$

c Draw $y = x + 2$ *(1 mark)*

$x = -2; x = 4$ *(1 mark)*

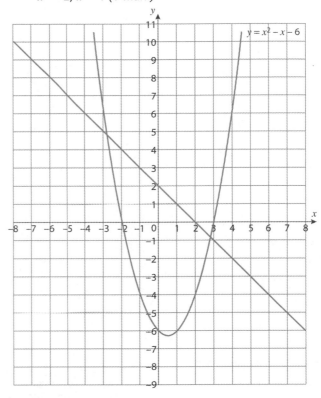

2 a −3.5, 0, 3.5 *(1 mark)*

b Draw $y = x$

−3.3, 0, 3.3 *(1 mark)*

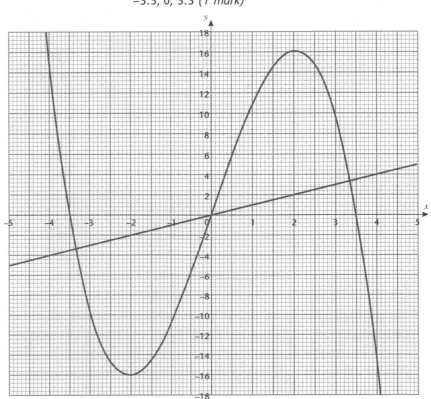

Page 175 Other graphs

1 a 0.125; 0.0625

b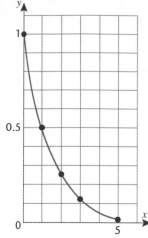

c 0.3

2 A (–4, 0)

B (–2, 0)

C (4, 0)

D (0, –32)

Page 176 Algebraic fractions

1 a $\dfrac{12x + 7}{6}$ *(1 mark for numerator, 1 mark for denominator)*

b $\dfrac{-4x + 23}{10}$

c $\dfrac{8x^2 - 6x + 1}{6}$

d $\dfrac{2x - 2}{3x - 1}$

2 a 1   *(1 mark for $\dfrac{2x + 3}{5} \times \dfrac{15}{6x + 9}$)*

b $\dfrac{2(x^2 - 3y^2)}{9}$ *(1 mark for $\dfrac{2x^2}{9} \times \dfrac{6y^2}{9}$)*

Page 177 Solving equations

1 a $x = 3$ *(1 mark for 6x = 18)*

b $x = 2$ *(1 mark for 11x = 22)*

c $x = -\frac{1}{9}$ or *1 mark for* $(9x + 1)(x - 1) = 0$

2 a *1 mark for* $3(3x - 1) - 4(2x - 1)$
= $(2x - 1)(3x - 1)$

1 mark for $x + 1 = 6x^2 - 5x + 1$

b $x = 0$ or 1

3 $x = \frac{1}{2}$ *(1 mark for $x^2 + x + 3x - 3 = x^2 - 1$)*

Page 178 Simultaneous equations 2

1 *1 mark* for $x^2 - x - 12 = 0$

1 mark for $(x - 4)(x + 3) = 0$

1 mark for (4, 0) or (–3, 7)

2 a $y = \frac{3}{2}x + \frac{13}{2}$

b (4, 7) and (–1, –8)   *(1 mark for $x^2 - 6x - 7 = 0$; 1 mark for $(x + 1)(x - 7) = 0$)*

3 (–3, –1) and (3, 1)   *(1 mark each; 1 mark for $11y^2 = 11$)*

Page 179 The nth term

1 a 5, 9, 13

b 7th

c Not odd

2 $7n - 4$ *(1 mark each term)*

3 a Always odd

b Always even

c Could be either

d Always odd

e Always odd

4 a 2 × anything is even

b $2n \times 2n = 4n^2$ *(1 mark)* which is a multiple of 4 *(1 mark)*

Page 180 Formulae

1 a 16, 21, 26

b 101

c $5n + 1$ *(1 mark each term)*

2 $x = \dfrac{2 + 4y}{y - 1}$ *(1 mark for $xy - x = 2 + 4y$)*

3 $6x + 2y = 4x + 12$ *(1 mark)*

$2x = 12 - 2y$ *(1 mark)*

$x = 6 - y$ *(1 mark)*

Page 181 Inequalities

1 a $x \leq 2$

   b $x > -2$

   c $-1, 0, 1, 2$

2 a $-3 < x \leq 1$

   b i $\frac{x}{2} > -2, x > -4$

   ii $x + 3 \leq 2$ *(1 mark)*; $x \leq -1$

   c $-3, -2, -1$

3 a $x \geq 4\frac{1}{2}$ *(1 mark for $2x \leq 9$)*

   b $x < -1$ *(1 mark for $2x < -2$)*

Page 182 Graphical inequalities

1 $x \geq -4, y \leq 3$ and $y \geq x$

2

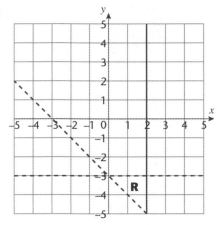

Page 183 Graph transforms

1 a

   b

   c

Page 184 Graph transforms

1 a

b

c

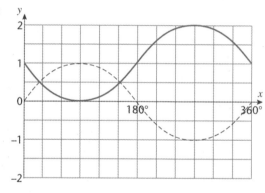

Page 185 Proof

1 $n^2 + 10n + 25 - (n^2 + 6n + 9)$ *(1 mark)*

    $= 4n + 16$ *(1 mark)*

    $= 4(n + 4)$ *(1 mark)*

2 *There are many ways to prove this but some angles will need to be found, 2 marks for 2 of these with reasons*

    CDB = 75° (alternate segment)

    CBD = 35° (alternates segment of angles in a triangle)

    DAB = 110° (opposite angles in cyclic quad)

    ADB = ABD = 35° (isosceles triangle)

* (1 mark for a reason why AD parallel to BC, e.g. ADC + DCB = 180°; ADB = DBC (alternate angles))*

3 $8 \times \frac{1}{2}n(n + 1) + 1 = 4n(n + 1) + 1$ *(1 mark)*

    $= 4n^2 + 4n + 1$

    $= (2n + 1)(2n + 1)$

    $= (2n + 1)^2$ *(1 mark)*